Authentication of Embedded Devices

Basel Halak

Editor

Authentication of Embedded Devices

Technologies, Protocols and Emerging Applications

 Springer

Editor
Basel Halak
University of Southampton
Southampton, UK

ISBN 978-3-030-60771-5   ISBN 978-3-030-60769-2   (eBook)
https://doi.org/10.1007/978-3-030-60769-2

This Springer imprint is published by the registered company Springer Nature Switzerland AG
The registered company address is: Gewerbestrasse 11, 6330 Cham, Switzerland

To
Suzanne, Hanin and Sophia
with Love.

Preface

Authentication is the process of verifying the identity of a person or a physical object. One of the earliest techniques for authentication dates back to the Babylonian time in 200 BC, wherein fingerprints were used to sign contracts, which are subsequently verified by visual inspection.

Nowadays, authentication has become an integral part of electronic engineering and computer science fields. A trend driven by the proliferation of computing devices in all corners of modern life and the increased reliance on these to access services and resources online. Additionally, there are a rising number of applications that require the verification of the identity of an electronic device, such as tracking goods in manufacturing processes; secure products transfer in the consumer services industry, gaining access to subscribed TV channels, cash withdrawal in banking systems, and boarder control using e-passports.

Each of these applications has a diverse set of security requirements and a different amount of resources, which means existing solutions are no longer sufficient, either because they are too weak or because they are too expensive. This has led to a number of successful large scale cyber-attacks on vulnerable devices such as those extensively used for the internet-of things applications.

Another factor that is driving the development of new authentication schemes the significant levels of outsourcing in the hardware supply chain, which has increased the risk of integrated circuits (IC) counterfeit and intellectual property (IP) piracy. This means that there is an increase in the need to verify the authenticity of the electronic circuity in order to ensure that computing device is not forged or compromised.

The development of secure and efficient authentication schemes is therefore crucial to ensure the continued safe use of digital technology.

The prime objective of this book is to provide a timely and coherent account of the latest advances in the key research areas of authentication technology for semiconductor products. It has been developed as a collaborative effort among two international research groups, each providing an-up-to date summary of their latest findings and highlighting remaining challenges and research opportunities. To facilitate the understanding of the material, each chapter includes background information explaining related terminologies and principles, in addition to a comprehensive list of relevant references. The book is divided into three parts to enhance its readability, namely; fingerprinting technologies, protocols and emerging applications of authentication schemes.

The Contents at Glance

This book explains principles of fingerprinting techniques with a special focus on hardware-based technologies, including physically unclonable functions and IC fingerprinting methods.

Afterword's, the book presents a number of state-of-the-art authentication protocols tailored for internet-of-things devices and energy-constrained systems. The development process of each protocol is discussed in details to allow reproducibility of the work. More specifically, the book explains how the design specifications of each protocol are derived from the system requirement, how to validate security assumptions using Scyther, and how to develop a hardware proof of concept for cost estimation and comparative analysis with existing solutions.

The book also discusses emerging applications for hardware-based authentication schemes, these include counterfeit mitigation techniques for the IC supply chain and anti-spoofing methods for the global positioning (GPS) system. More details on each chapter are provided below.

Part I: Fingerprinting Technologies.

Chapter 1 provides a comprehensive review of the fingerprinting technologies, with special, focus on circuit-based techniques.
Chapter 2 discusses in details the principles of physically unclonable functions (PUF), including design metrics, constructions and main applications.

Part II: Authentication Protocols.

Chapter 3 presents an authentication protocol and a key agreement scheme for the internet of things devices, which is based on the use of PUF technology to provide better physical security and more energy efficiency.
Chapter 4 present a two flight authentication protocol for energy constrained systems, which is based on combining elliptic curve cryptography with the use of a lightweight symmetric cipher.

Part III: Emerging Applications of Hardware-based Authentication.

Chapter 5 presents a hardware-based approach for products' authentication and tracking to mitigate the risk of counterfeiting in the IC supply chain. The technique is based on consortium blockchain and smart contract technologies, wherein each device incorporate a PUF that generates its unique digital identity.

Chapter 6 discusses hardware-oriented security applications for the authentication of users, devices, and data, and illustrates how physical properties of computing hardware (e.g. memory, computing units, and clocks) can be used for authentication applications in low power devices and the global positioning system (GPS).

Book Audience

The book is intended to provide a comprehensive coverage of the latest research advances in the key research areas of authentication technologies and protocols; this makes it a valuable resource for graduate students researchers, and engineers working in these areas. I hope this book will complement the ongoing research and teaching activities in this field.

Southampton, UK   Basel Halak
June, 2020

Acknowledgments

I would like to thank all of those who contributed to the emergence, creation and correction of this book.

Firstly, I gratefully acknowledge the valuable contributions from all the authors, for taking the time to share his knowledge and for being very accommodating throughout the publication process. Special thanks go to the graduate students at the University of Southampton, University of Maryland and Morgan State University for the many hours they have spent working in their labs to generate the experimental results. Of course, the book would not be successful without the contributions of many researches and expert in hardware security and authentication protocols.

Finally, I would like to thank the great team at Springer for their help and support throughout the publication process.

Contents

About the Editor

Basel Halak is the director of the embedded systems and IoT program at the University of Southampton, a visiting scholar at the Technical University of Kaiserslautern, a visiting professor at the Kazakh-British Technical University, an industrial fellow of the royal academy of engineering and a senior fellow of the higher education academy. He has written over 70-refereed conference and journal papers, and authored four books, including the first textbook on Physically Unclonable Functions. His research expertise include evaluation of security of hardware devices, development of appropriate countermeasures, the development of mathematical formalisms of reliability issues in CMOS circuits (e.g. crosstalk, radiation, ageing), and the use of fault tolerance techniques to improve the robustness of electronics systems against such issues.. Dr. Halak lectures on digital design, Secure Hardware and Cryptography, supervises a number of MSc and PhD students, and is the ECS Exchange Coordinators. He is also leading European Masters in Embedded Computing Systems (EMECS), a 2 year course run in collaboration with Kaiserslautern University in Germany and the Norwegian University of Science and Technology in Trondheim (electronics and communication). Dr. Halak serves on several technical program committees such as HOST, IEEE DATE, IVSW, ICCCA, ICCCS, MTV and EWME. He is an associate editor of IEEE access and an editor of the IET circuit devices and system journal. He is also a member of the hardware security-working group of the World Wide Web Consortium (W3C).

Part I
Fingerprinting Technologies

Chapter 1
Integrated Circuit Digital Fingerprinting–Based Authentication

Xi Chen and Gang Qu

Abstract As we move to the era of the Internet of Things (IoT), the embedded devices in IoT applications may contain a lot of sensitive information and many of them are attached to humans. This makes security and trust of these devices a new and challenging design objective. Device authentication is critical for any security-related features, but current cryptography-based authentication protocols are computational expensive. In this chapter, after a brief introduction of hardware-based lightweight authentication for embedded devices, we will focus on integrated circuit (IC) fingerprinting. Like human fingerprints, which have been used for thousands of years for identification, the key idea behind IC fingerprinting is to extract certain unique physical characteristics from the IC such that they can be used to identify and authenticate the chip (or device).

Digital fingerprinting was first proposed in 1999 for the protection of very large scale integration (VLSI) design intellectual properties (IP). Various techniques have been developed to make each copy of the IP unique in order to track the usage of the IP and trace any traitors who have misused the IP. We will review the general requirements and the available schemes to create digital fingerprints for IP protection. We will then discuss the challenges of applying these methods for device authentication in IoT applications and how to overcome these difficulties. As an example, we consider the fact that embedded devices are designed by reusing IP cores with reconfigurable scan network (RSN) as the standard testing facility and elaborate how to generate unique IC identifications (IDs) based on different configurations for the RSN. These circuit IDs can be used as IC fingerprints to solve the device identification and authentication problems. This IC fingerprinting method complies with the IEEE standards and thus has a high practical value.

Keywords IC Fingerprinting · Intellectual Properties (IP) · VLSI · Device authentication · IoT · Reconfigurable Scan Network (RSN)

X. Chen · G. Qu (✉)
University of Maryland, College Park, MD, USA
e-mail: xichen128@umd.edu; gangqu@umd.edu

© Springer Nature Switzerland AG 2021
B. Halak (ed.), *Authentication of Embedded Devices*,
https://doi.org/10.1007/978-3-030-60769-2_1

1.1 Introduction

An embedded system is a combination of (dedicated) software running on (customized) hardware and memory with application-specific input/output peripheral devices within another larger electrical or mechanical system. Although it is generally believed that the first embedded system is MIT's Apollo Guidance Computer designed to collect data at real time and to perform mission- and time-critical calculations for the Apollo Program, this concept was quickly picked up by the automobile industry and military applications for real-time computation, command control, and communications. In the late 1990s, with the advances in the Internet, wireless communications, and semiconductor fabrication technologies, the networked miniature embedded systems became ubiquitous and found all sorts of applications. Popular embedded systems can be found in multimedia applications (digital cameras, camcorders, TV set-top boxes, DVD players) and people's daily life (cell phones, answering machines, toasters, personal digital assistants). The key challenges for the design of these embedded systems were cost, power, size, safety, and time-to-market.

Then we entered the era of the Internet of Things (IoT). IoT is a group of devices or embedded systems that are connected by the Internet infrastructure to accomplish one or more specific applications without human interaction. Examples of the IoT include medical and healthcare systems, smart homes and buildings, and large nation-wide infrastructures such as power grid, transportation systems, and environmental monitoring systems. The embedded devices or the *THINGS* in the IoT normally have the capabilities of sensing (to collect data), computing (to process data and obtain knowledge), communication (to exchange data and knowledge), and execution (to carry out actions based on the knowledge). These *THINGS* can be sensors deployed for wild fire, earthquake, or landslide monitoring; they can be a Wi-Fi-enabled pacemaker, a smart meter in the power grid, or tire pressure sensors for an automobile. These IoT applications have brought, in addition to the design challenges for the traditional embedded systems, a set of new design objectives including: trust, security, privacy, ultra-low power, safety, and reliability [1]. The focus of this book is one important aspect of these challenges – the authentication of embedded devices, where we will review the current technologies and protocols for authentication with illustrations on emerging applications.

Embedded devices play a vital role in IoT applications; therefore, the authentication of these devices is equally important. The authenticity of a device not only identifies the device for system safety and maintenance, it also helps in cross-validating the integrity of the collected data, identifying the larger system that the device is embedded in and the user of the system, and enhancing system overall security. On the other hand, a compromised device could forge data, maliciously produce faulty results to mislead the decision-making process, or leak sensitive information of the user of the device. As we have pointed out in [1], utilizing hardware characteristics in the embedded devices for authentication has several advantages over traditional cryptographic solutions. In this chapter, we will discuss digital fingerprinting-based approaches for device authentication with the focus on integrated circuit fingerprinting. Chapter 2 is dedicated to authentication based on physical unclonable function (PUF).

1.2 Chapter Overview

We aim to provide a comprehensive picture of the research and development of hardware digital fingerprint in the field of electronic design automation (EDA) community and the more general embedded systems design society. As we have introduced in the previous section, we will emphasize our discussion in digital fingerprinting–based device authentication.

First, in Sect. 1.3, we discuss the importance of authentication in embedded devices and IoT in general. We then briefly review some popular cryptographic solutions for authentication, which are computational expensive and thus are not applicable to the resource-constrained IoT devices. This motivates us to use hardware for lightweight authentication, which has some intrinsic advantages and is promising for IoT applications.

Then, in Sect. 1.4, we provide the foundations for integrated circuit (IC) fingerprinting-based authentication. We start with a review of the history of digital fingerprinting for VLSI design intellectual property protection. We then present the requirements and the general principles of using IC fingerprint for device authentication.

Next, in Sect. 1.5, we survey the representative approaches of IC fingerprinting techniques. These include both pre-silicon and post-silicon methods that cover all phases of IC design, fabrication, and testing.

We elaborate the details of a recently proposed circuit fingerprinting scheme based on reconfigurable scan chain network in Sect. 1.6. Because of the facts that scan chain and scan network are implemented in modern designs and the proposed scheme complies with IEEE standards, it has high practical value. Section 1.7 summarizes this chapter.

1.3 Hardware-Based Lightweight Authentication

1.3.1 The Need of Authentication in Embedded Devices

In 2013, the telecommunication giant Cisco Systems, Inc. introduced the term Internet of Everything (IoE), which brings "*together people, process, data, and things to make networked connections more relevant and valuable than ever before-turning information into actions that create new capabilities, richer experiences, and unprecedented economic opportunity for businesses, individuals, and coun-*

tries". While IoE emphasizes the connection among the four pillars: people, process, data, and things, the things or embedded devices serve as its root. Data need to be collected and processed by devices such as sensors, people rely on devices such as smart phones to stay connected, and computer hardware does the process.

The goal of process in IoE is to "*deliver the right information to the right person (or machine) at the right time*". Authentication is the nature answer to these *rights* for all the four pillars of IoE. Although the *right* device is not explicitly mentioned in this statement, device authentication is the foundation for the authentication of others. The right information has to be collected by the right and trusted devices to ensure that the raw data is right to start with. The data integrity check is needed to verify that the data has not been modified during data transmission. On top of integrity check, devices at the senders, receivers, and all the nodes in between must be authenticated for data security. The right person can be authenticated with the person's biometric information or a password, as we will elaborate in the next section. Dedicated devices are required for such authentication and it is obvious that such devices must be authenticated and trusted. The right time can be an absolute time, which normally uses GPS as a reference, or a time instant depending on the occurrence of other events. In Chap. 6, we detail one novel approach that utilizes the system clock in the local embedded device to cross-validate the GPS signals in order to detect GPS spoofing attacks. As a conclusion, one can see that device authentication is the key for IoE security.

In 2014, when the EDA perspectives of embedded device design for IoT were presented [1], the authors listed the following design challenges for security and privacy:

- Which *THINGS* are collecting data/information?
- What data/information is collected by the *THINGS*?
- How data/information is collected and stored?
- Whom will the *THINGS* share the data/information with?
- How data/information is communicated among the *THINGS* and others?

These questions coincide precisely with the specific tasks to verify for the process we discussed above. Similarly, the answer to all of these questions is authentication. One needs to authenticate the device to know which *THINGS* are collecting data, to authenticate the data and the process of data collection and storage to ensure data integrity, to authenticate users and other devices that the *THINGS* are interacting with, and to authenticate the communication channel.

1.3.2 Classical Authentication Protocols

We briefly discuss digital signature and challenge-response protocol for authentication. More details can be found in Part II of this book. The goal of this introduction to authentication protocols is to show their high computation and run time complexities and to give the motivation for hardware-based lightweight authentication.

In the context of digital data authentication, digital signature and challenge-response protocol are two of the most popular methods. Digital signature can authenticate both the source of the message and its integrity (which means that the message has not been altered). A digital signature scheme normally has three components: key generation, which selects the cryptographic keys; signature signing, which creates the signature for a given message using the private key; and signature verifying, which validates the authenticity of the signature and the message using the public key. Current digital signature schemes in use include digital signature algorithm (DSA), elliptic curve DSA, RSA-based signatures, and so on. These schemes are all based on intractable mathematical problems and involve expensive computations such as modular exponentiation where the exponents are cryptographic keys. As it is recommended to use 1024-bit or at least 512-bit keys, most of the embedded devices cannot afford to implement traditional digital signature–based authentication.

In the challenge-response authentication protocol, one party (the verifier) asks a question (i.e., challenge) and the other party (the proofer) must provide an answer (i.e., response), which the verifier will check and then decide to accept or reject the proofer's authenticity claim. One of the simplest examples of challenge-response protocol is the password authentication method, which many systems are using. In such a method, a user (the proofer) will provide a pair of user name and password and the system will then verify the password based on the user name to authenticate the user (actually the pair of user name and password). Validation of the response typically relies on some cryptographic operation. The responses need to be stored in the system securely. Both will be a challenge for the resource-constrained embedded devices.

Recently, there is the trend of using multifactor authentications, which requires two or more of the following factors: knowledge factors (such as password, challenge-response pair, or answers to security questions that the user knows), ownership factors (such as a smart phone or an ID card that the user has), inherence factors (such as the fingerprint, iris, voice, or movement that are unique and can be used to identify the user), and other factors (such as the location where the user is). Note that most of the multifactor authentication protocols are a combination of cryptographic based methods and other noncrypto mechanisms. Although enhancing security is the goal of such combination, we see that the computation complexity is reduced at the same time.

1.3.3   Hardware-Based Lightweight Authentication

In the above multifactor authentication protocols, the noncrypto mechanisms take advantages of certain physical objects that belong to the user for authentication. For example, a code sent to the user's smart phone can be used to verify that the claimed user does possess the phone. The fingerprint or iris information of the user can be matched to that stored in the database to authenticate the user.

These operations normally do not necessarily require cryptographic computation. However, they lack a sound mathematical foundation to prove the security level of the authentication they can provide. So in multifactor authentication, these are considered as the enhancement of the crypto-based knowledge factors.

Security and privacy are among the key concerns for the development of IoT applications and the design of the IoT devices. A January 2014 article in Forbes listed many Internet-connected appliances that can already "spy on people in their own homes," including televisions, kitchen appliances, cameras, and thermostats [2]. Embedded devices in automobiles such as brakes, engine, locks, hood and truck releases, horn, heat, and dashboard have been shown to be vulnerable to attackers who have access to the on-board network. The vehicle-to-vehicle and vehicle-to-infrastructure communication makes everyone's driving habit and daily commute route public [3].

Mathematically strong and well-developed cryptographic techniques exist for all kinds of security-related applications such as data encryption/decryption, user and devices authentication, secure computation and communication. Most of these crypto security primitives or protocols are (extremely) computationally expensive (for example, performing the modular exponentiation operation for large numbers of hundreds of bits). Unfortunately, in the IoT domain, the devices are extremely resource constrained and do not have the required computational power, memory, or (battery) power for such operations. As a result, in many IoT applications, both data and control communications, such as those between wearable/implantable medical devices and doctor or patient, are in plain text, which creates serious security vulnerabilities.

Authentication based on physical features (such as bioinformatics and hardware manufacture variations) is built on the following two important observations in IoT applications and embedded devices:

Constrained resources. The computation power, memory size, CPU speed, battery capacity, communication bandwidth, transmission range, and other resources on many embedded devices are limited and hard or impossible to renew. Thus, the mathematical sound cryptographic solutions are not applicable.

Imperfect match. Cryptographic algorithms work in the digital domain, where even one bit of error is considered as a mismatch and an erroneous bit in the key could alter many bits. However, the IoT applications normally deal with analog world where a perfect match, for example, in human fingerprint or voice, is hard to find and not necessary for authentication purpose.

Embedded device authentication demands resource efficiency but not the highest level of security. This gives us the opportunity to develop noncryptographic solutions. One promising direction is using the hardware (i.e., the chips or ICs) in the embedded devices. In this chapter, we will focus on how to generate IC fingerprints for such purpose.

1.4 Principles of Digital Fingerprinting–Based Device Authentication

1.4.1 IC Digital Fingerprints

Fingerprints are the characteristic of an object that is completely unique and incontrovertible so they can be used to identify a particular object from its peers. They have been used for human identification for ages and have been adopted in multimedia for copyright protection of widely distributed digital data. In the semiconductor and IC industry, the concept of digital fingerprinting was proposed in the late 1990s with the goal of protecting design IP from being misused [4–8]. In this context, digital fingerprints refer to *additional features that are embedded during the design and fabrication process to make each copy of the design unique*. These features can be extracted from the IP to establish the fingerprint for the purposes of identification and protection.

In [9], a new semiconductor capacitive sensor structure is proposed and the fabrication process for a single-chip fingerprint sensor stacked on a 0.5-/spl mu/m CMOS LSI is presented. This technology has been successfully commercialized, and it currently used in IoT devices for identification purpose. In [10, 11], the concept of IP metering was proposed to allow design houses to achieve postfabrication control over their ICs. The enabling technology behind metering is the creation of unique identifiers (also called tags or fingerprints) for each copy of the IP. A comprehensive survey on hardware metering can be found in [12], and we will not elaborate here. Several practical IP fingerprinting techniques were studied in [13–15], which we will discuss in the next section.

In the context of IP protection by fingerprinting, the goal is to give each user a copy of the IP containing a unique fingerprint, which can be used to identify that user in order to prove that he/she is innocent should an illegally copied IP being found. It is one instance of the so-called *statistical fingerprinting* that can be characterized as: given sufficiently many misused objects to examine, the distributor can gain any desired degree of confidence that he/she has correctly identified the compromised [16]. The identification of the fingerprint is, however, never certain. When hardware-based fingerprint is used for device authentication, this principle also applies. It is because of this fact that there is no theoretical guarantee on the strength of authentication, we refer to this approach as *lightweight authentication*.

1.4.2 Requirements for Hardware-Based Device Authentication

Based on the requirements for a fingerprinting scheme to be effective [15] and considering the specialties in embedded device authentication, we propose the following requirements for fingerprinting-based device authentication:

High credibility. The fingerprint or other hardware identifiers should be readily detectable in proving the claimed identity for authentication. The probability of coincidence (i.e., the fingerprint is caused by accident, not by design) should be low.

Low cost. The hardware-based authentication process should incur minimal resource of all kinds, including memory, energy consumption, and communication bandwidth. If the fingerprints are added only for authentication, ideally it should not introduce any design overhead.

Fast processing. The verification process of the fingerprint should be sufficiently fast in order to satisfy the requirement of real-time device authentication for many of the application scenarios.

High resilience. The fingerprint and other hardware features used for authentication purpose should be difficult or impossible to remove even with partial or complete knowledge of the authentication protocol.

No information leak. The fingerprint and other hardware features are designed to verify device identity. They should not leak any other sensitive information about the device and the data it collects or stores during the authentication process.

Large volume. Because each device must have a unique fingerprint or other hardware feature, there should be abundant of them to accommodate the large amount of embedded devices deployed in various applications. The run-time for creating these fingerprints in bulk must be low as it is impractical to generate them one by one.

High reliability. The fingerprint or other hardware feature should be ready to be extracted and processed for device authentication under the working environment of the devices. Due to the variety of the devices and the harsh environment they might be deployed to, these hardware identifiers must be reliable at the variation of factors such as temperature, humidity, altitude, air pressure, radiation, device aging, etc., which will be highly application dependent.

Collusion free. Identical devices, although they are identical, should receive different hardware identifiers or fingerprints to distinguish them from each other. These identifiers or fingerprints should be designed in such a way that it is difficult to forge a new identifier or fingerprint from the existing ones.

Most of the above requirements are not restricted to fingerprint, they are applicable to device authentication methods based on other hardware features as well. For example, silicon physical unclonable function (PUF) is considered as a unique intrinsic hardware characteristic and Chap. 2 of this book is dedicated to the authentication with PUF. As we will see in that chapter, the design and implementation of PUF based device fingerprint also consider such requirements.

1.4.3   Two-Phase Hardware-Based Device Authentication

Like other authentication protocols, hardware-based device authentication methods also have two phases: *registration* and *verification*, which we will elaborate with the example of digital fingerprint. Registration phase is the process of enrolling the fingerprint and the object (human, device, etc.) it belongs to in a database. Verification phase is the process of extracting the fingerprint from an object and determining whether it matches the fingerprint stored in the database that belongs to the claimed object. If a match is found, the object is considered to be authenticated. Otherwise, the authentication fails.

The general assumption behind fingerprint-based human authentication method is that every person has a unique fingerprint. Although there is no scientific proof, this seems to be true as it has never been reported that two people, even identical twins, have exactly the same fingerprints. This assumption is clearly necessary because if multiple people have the same fingerprints, it will become impossible to distinguish them from the fingerprints, let alone authenticating anyone. Therefore, any fingerprint for authentication purpose must be based on hardware features that are known to be unique. For example, silicon PUF fingerprint (which is elaborated in Chap. 2) relies on the belief that semiconductor manufacture variation is random, unclonable, and unpredictable. Unlike the intrinsic silicon PUF, we will discuss how to create IC fingerprints for a device in the next two sections of this chapter. One advantage of these man-made IC fingerprints is that they are guaranteed to be unique.

Fingerprint registration is the next step after fingerprints are collected. Without registering the fingerprint and the device that it belongs to in a database, one can only draw negative conclusions, such as that the fingerprint does not belong to a certain device or that the two fingerprints are different. Apparently this will not suffice for the purpose of device authentication. How to design and maintain a secure fingerprint database is a well-studied question in database. It is out of our scope and there are many available solutions.

Once a device's fingerprint is registered in a secure and trusted database, one can verify the fingerprint in order to authenticate the device. Needless to say, this verification process must be performed in a secure, accurate, and time-efficient way. We can adopt existing verification protocols developed for other digital contents. Our challenge is how to collect or extract the hardware fingerprint from the device. This is highly dependent on the fingerprint creation method. In the next two sections, we will discuss some of the representative IC fingerprinting techniques, with the emphasis on how digital fingerprints can be embedded in all stages of IC design.

To end this section, we mention some of the lessons we have learned from human fingerprinting authentication. They are valuable for the design of device authentication schemes based on (analog) hardware features.

IC fingerprints and other hardware features, like human fingerprints, are analog by nature. However, when they are captured and registered for authentication, they will be converted into digital format. For instance, Fig. 1.1 is the fingerprint used

Fig. 1.1 The fingerprint used
in a US legal document

in a US legal document. The image of a fingerprint is scanned and embedded in the card. This image will be used to authenticate the owner of the fingerprint. What differentiates it from the real human fingerprint is that this is a static image with limited information. This difference creates several challenges for authentication.

First, it is the loss of information during the analog to digital conversion. No matter how precise the image could be, or how many bits we use to represent the image, it will not be the complete information of the analog fingerprint, which creates the fault positive when a different fingerprint matches the static image. This problem could be serious for embedded IoT devices as they are resource constrained and the precision of the collected fingerprint will be limited.

Secondly, it is the loss of authentication accuracy. Although it is believed that a person's fingerprint will not change through his/her lifetime, the fingerprint collected for authentication may not be exactly the same as the static image stored in the database due to various factors such as the condition of the skin on the finger, the pressure applied on the collecting device by the finger, and the systematic bias of the collecting device. This would result in false negative and fail to authenticate the authentic person.

Thirdly, probabilistic verification is used in practice. A tiny mismatch in analog domain can still give a very high overall matching. But in digital domain, in particular when cryptographic solutions such as encryption are used, even one bit of difference can result in significant mismatch. Therefore, human fingerprint authentication never requires a perfect match. A significant portion of the match (e.g., 70% or above) will be considered as a success in authentication.

1.5 Practices in Integrated Circuit Fingerprinting

In this section, we briefly review representative approaches in creating digital fingerprints in various stages of IC design and fabrication. A detailed description of these approaches can be found in [15]. As we have discussed, we will focus on how to generate distinct digital fingerprints.

1.5.1   *Fingerprinting by Constraint Addition*

This method can be applied in any design stages. It requires the abstract modeling of the corresponding design problem and clever techniques to generate distinct solutions to the problem efficiently. In [7–8], this method is elaborated by the example of graph coloring problem. Here we illustrate the idea using the graph partitioning problem.

Given a graph, the graph partitioning problem seeks to partition the vertices into multiple nonempty disjoint subsets. The objectives of the partition vary from the real problems. Some of the commonly used ones are: *balanced partition*, which requires the number of vertices in each subset to be the same or close to each other; *minimized connection*, which requires the number of edges between the vertices that are partitioned to different subsets to be as few as possible. For example, in Fig. 1.2, the dashed line partitions the graph into two subsets, the top half and the bottom half. This partition is perfectly balanced as each half has exactly 12 vertices. It has 10 connections between the two subsets because it cuts through 10 edges.

Now in order to leave fingerprint in the solutions to this graph partition problem, we need to create distinct solutions. Figure 1.3 illustrates how an 8-bit fingerprint is embedded in a solution to make it unique. First 8 pairs of vertices are selected and labelled as {0 0}, {1 1}, ..., {7 7}. In a graph partition solution, the two vertices in each pair can be either in the same subset or in different subsets. We then convert the 8-bit fingerprint into additional constraints for the partition such that the partition solution will have specific features on these 8 pairs of vertices. This can be achieved as follows:

If the i-th bit in the fingerprint is 0, we require both vertices labelled with i to be in the same subset. This can be easily accomplished by assigning the two vertices in the same subset before we partition the graph.

If the i-th bit in the fingerprint is 1, we require the two vertices labelled with i to be partitioned in different subsets. This can also be easily achieved before we partition the graph.

Fig. 1.2 A graph with 24 vertices for the graph partitioning problem

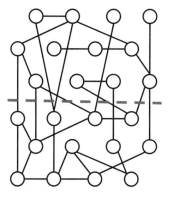

Fig. 1.3 Creating 8-bit
fingerprints on the solutions
to the graph partitioning
problem

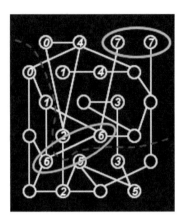

After these 8 pairs of vertices are partitioned, we can use any available tool to partition the remaining vertices in the graph to obtain a complete partition solution for all the vertices in the graph. The dashed curve in Fig. 1.3 shows a solution with a fingerprint. We now use this as an example to show how the fingerprint can be retrieved from this partition solution. According to the scheme that each bit in the fingerprint is embedded, we go over all the 8 pairs of vertices, starting from pair 7, to check whether each pair is separated in two subsets (which implies fingerprint bit 1) or remains in the same subset (which indicates fingerprint bit 0). This will give us the following 8-bit fingerprint: *01001111*. As a comparison, the 8-bit fingerprint for the partition in Fig. 1.2 is 00001000 because only the pair of vertices labelled with 3 are separated.

The advantage of this approach is that it guarantees that different fingerprints will always lead to different solutions. The fingerprint can be generated in early design stage in the form of special properties in the final design, which makes it simple to verify the fingerprint but hard to forge one without significant redesign efforts. However, one of its weaknesses makes it impractical. That is, after the fingerprint is embedded as special constraints, each fingerprinted design becomes a different design and has to be designed separately. In [7], a novel constrain-addition method was proposed to solve this problem. But the final design for each fingerprinted copy is still distinct. Technically each design will require a different mask for fabrication, making it extremely expensive, if not impossible, for fabrication.

1.5.2 Fingerprinting by Iterative Refinement

This\enlargethispage{-12pt} approach, first proposed in [5–6], is based on the fact that many hard problems in IC design are solved iteratively where an initial solution is first obtained and then refined to improve the quality repetitively until the solution becomes satisfiable. In each iteration, the current and previous solutions

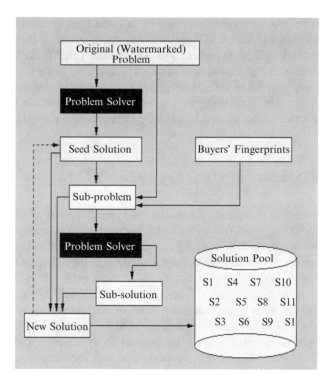

Fig. 1.4 Illustration of the iterative refinement fingerprinting approach

are analyzed carefully and additional conditionals could be added to enforce that the solution found next will always be new.

Inspired by this, iterative refinement fingerprinting approach was proposed to reduce the design time for the large amount of designs with different fingerprints. As shown in Fig. 1.4, starting with the original problem, the time-consuming computer aided design (CAD) tools or problem solver will be used to find a seed solution. This process normally takes long time in order to guarantee the high quality of the seed solution. Then from this seed solution and the fingerprint intended to be embedded for authentication, we build a subproblem of small size such that the current solutions violate certain requirements and thus become invalid for the subproblem. For example, in the graph partition problem, we can require two nodes that are currently in the same subset to be partitioned in different subsets, or vice versa. Next we use the CAD tools or problem solver again to find a solution to the subproblem, which can be integrated with the seed solution (or solutions from the previous iterations) to form a new solution to the original problem. This solution is new because it meets the deliberately added requirements that the previous solutions do not meet.

Specific fingerprinting techniques based on this idea were developed [5] for four important problems in VLSI CAD: partitioning, graph coloring, Boolean satisfiability, and standard-cell placement. It was demonstrated that these fingerprinting techniques are effective on a number of standard benchmarks in the tradeoff between runtime to create new solutions and resilience against collusion attacks. However, similar to the constraint-addition approach in the previous subsection, this approach may not be practical either when it is applied on the pre-silicon stages.

1.5.3   Post-Silicon Fingerprinting on Don't-Care Conditions

The constraint-addition and iterative refinement approaches are effective in generating large amounts of solutions with distinct hardware features, which provide the foundation for fingerprinting. But they are not practical because these techniques work on pre-silicon stages, and the fingerprinted designs they produce need separate masks for fabrication. This motivates us to develop post-silicon fingerprinting techniques where the fingerprints will be created after chip fabrication. This implies that the fabricated chips must have some kind of configurability, otherwise they will be all identical and cannot be authenticated by digital fingerprints (except manufacture variation–based fingerprints such as PUF, which is discussed in the next chapter). In this subsection, we introduce two methods to create digital fingerprints that can be easily verified. One difference between these fingerprints and silicon PUF is that they have better controllability.

1.5.3.1   Observability Don't Care–Based Fingerprinting

Observability don't care (ODC) conditions were utilized to create fingerprinting copies [14]. An ODC condition occurs when local signal changes cannot be observed at a primary output. For example, in Fig. 1.5, at gate $F = XY$, when $Y = 0$ (the ODC condition), F will be 0 regardless of the value of X.

When the ODC condition fails, that is, $Y = 1$, $X = AB$ on the left circuit and $X = AYB = AB$ on the right circuit. So these two circuits implement exactly the same function although they are different. This feature allows us to create a 1-bit

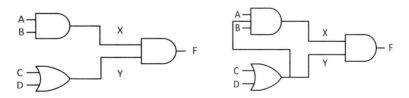

Fig. 1.5 Two 4-input circuits that implement the same function [14]

fingerprint by fabricating the right circuit with the flexibility of disconnecting the wire from Y to the AND gate that generates X (e.g., by using a fuse) at the post-silicon stage.

A practical ODC-based fingerprinting scheme is proposed based on this idea as follows: (1) a normal design is performed to get the best possible IP; (2) certain ODC conditions are identified and the circuit is modified locally to provide flexibility in implementing the same IP; (3) mask is built and identical circuits are fabricated; (4) fingerprinted copies are created at post-silicon stage by choosing whether or not to keep the local changes in these fingerprint locations.

Since ODC conditions are abundant in all combinational circuits, one will be able to find sufficient fingerprint locations to embed fingerprints. Furthermore, the changes made on the circuit in step (2) are minute; thus, it will cause small overhead in performance. The area and power increase introduced by the scheme were acceptable according to the simulation results. However, large and unacceptable delay overhead may occur. This is because the fingerprinting scheme didn't check whether circuit changes are made on the critical paths with tight timing constraint.

1.5.3.2   Satisfiability Don't Care–Based Fingerprinting

A satisfiability don't care (SDC) condition–based IC fingerprinting scheme was proposed in [13]. SDCs are the values of certain signals that will never occur because of their logical dependence. Consider the 2-input NAND gate in Fig. 1.6, none of the followings can happen: $\{A = 0, C = 0\}$, $\{B = 0, C = 0\}$ or $\{A = 1, B = 1, C = 1\}$, and they are SDC conditions.

Clearly the only difference between the two circuits in Fig. 1.6 (a) is the logic gate that generates the output: a 2-input NAND on the left circuit and a 2-input XOR gate on the right. However, the truth tables for the two circuits in Fig. 1.6 (b) shows that these two circuits have identical output signals for every input combination. This is because an NAND gate and an XOR gate generate different outputs only

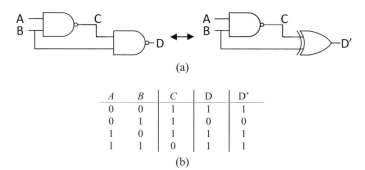

(a)

A	B	C	D	D'
0	0	1	1	1
0	1	1	0	0
1	0	1	1	1
1	1	0	1	1

(b)

Fig. 1.6 SDC-based fingerprinting [13]. (**a**) Two gate level circuits (**b**) Truth table for the two circuits

when both input signals B and C are 0, which is an SDC condition. So the work in [13] utilized this feature of circuits to encode one bit of fingerprinting information by deliberately choosing which gate to use to implement the circuit.

By locating gates that have SDCs leading into them, referred to as fingerprint locations, and finding alternative gates, different fingerprinted copies can be generated by using either the original gate or one of its alternatives at each fingerprint location. Two methods were proposed to replace the original gate, using either another library gate where the two gates have different outputs only on the SDC conditions or a $2^m \times 1$ multiplexer (MUX) since it can realize any m-input function. The MUX is preferred due to its flexibility for post-silicon configuration, which can bypass the need for expensive redesign for each fingerprint copy. However, MUX itself is much larger and slower than standard gates, which results in significant overhead in circuits. Hence, performance-driven methods need to be developed to control the overhead for this scheme.

1.5.4 Scan Chain–Based Fingerprinting

Both ODC- and SDC-based fingerprinting techniques introduce fingerprints at the post-silicon stage to reduce fabrication overhead. However, these two approaches entail large performance overhead and lack effective methods to verify the fingerprints. In [17], a scheme with easy fingerprint detection and low overhead was presented to leverage the controllability and observability of scan chain. As depicted in Fig. 1.7, five scan flip flops (SFFs) are connected in sequence to form a scan chain. A scan in (SI) port is available to shift test vectors into the scan chain, which allows test engineers to put the core under test (CUT) into any state for test. After testing, contents of the SFFs can be read out from the scan out (SO) port and then compared with the expected values to check whether the CUT behaves correctly.

Several previous works [18–19] have reported that scan cells can be chained by either the Q-SD or the Q'-SD connection style. Built on top of this, the proposed scheme embedded fingerprints by altering the connection styles between SFFs. In

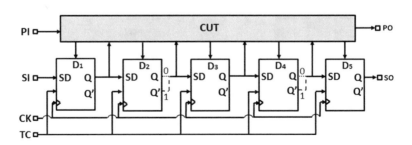

Fig. 1.7 Scan chain–based fingerprinting [17]. A 5-bit scan chain with the second and fourth connections chosen as the fingerprinting locations

Fig. 1.5, two pairs of SFFs, (D2, D3) and (D4, D5), have been identified as the locations to embed the fingerprint. The approach utilizes the Q-SD connection to embed a bit '0' and the Q'-SD connection as a bit '1'. As a result, this can embed any 2-bit fingerprint, "00," "01," "10," or "11," by selecting different connection styles in these two locations. To maintain the test vectors' fault coverage, the set of test vectors have to be updated accordingly. Hence, the fingerprint can be easily detected by checking the corresponding test vectors.

The fingerprinting process is as follows: (1) perform the normal scan design to obtain the best possible solution; (2) identify the fingerprint locations and create a bit of information at each fingerprint location by deliberately choosing whether two adjacent flip-flops have a Q-SD or a Q'-SD connection; (3) develop fingerprint embedding protocols; (4) modify the set of test vectors. The overhead caused by this scheme is minimal as its only effect is to increase the power usage of the device during testing. The approach is an ideal solution to fingerprinting circuits that utilize scan chains for design for test (DfT).

1.6   A Reconfigurable Scan Network–Based Circuit Fingerprint for Device Authentication

In this section, we propose a practical IC fingerprinting method which can be considered as a hardware security primitive for the security of embedded and IoT devices. We utilize the testing infrastructure in these devices, which is compliant with IEEE 1149.1-2013 [20] and IEEE P1687 (IJTAG) [21], to create unique identifier at the circuit level for each device which can be verified through standard testing interface. More specifically, we adopt the reconfigurable scan network (RSN) and develop a fingerprint protocol to configure distinct RSN for each IC by utilizing the different connection styles between scan flip flops. The testing vector set will need to be modified consequently to reflect the different RSN configurations and thus can be used as IC identification (ID). In addition, these IDs can be used to fingerprint the design or the embedded device, they can facilitate IP metering and tracking, they can also be used as the key for lightweight encryption and decryption.

1.6.1   Reconfigurable Scan Networks

Scan chains are extensively used to reduce the test complexity. They eliminate the need for sequential test pattern generation by making internal memory elements directly controllable and observable. However, in the traditional design of scan chains, where all scan registers are chained into a single scan chain, the time overhead of accessing each module's scan register can be too high. To reduce this overhead, reconfigurable scan networks (RSN) are introduced, which enable dynamic reconfiguration of scan networks and allow cost-efficient access to on-chip instrumentations.

An RSN has four data ports, namely, *scan-input*, *scan-output*, *reset input*, and *clock input*, as well as three control ports, *capture, shift, and update*, which are controlled by a 1149.1-compliant TAP [20]. RSNs are composed of *scan segments*, multiplexers, or other combinational logic blocks. The scan segment consists of scan registers which are accessible through the scan-in and scan-out ports and an optional shadow register. The state of the shadow register determines the configuration of scan networks. Scan segments provide access to testing structures and enable distributed control over the on-chip instrumentations. Each scan segment should support three modes of operations, namely, *shift, capture,* and *update*, which are controlled by external control signals.

In the capture mode, the scan registers get overwritten by the data coming from the corresponding instrument (data-in port). During a shift operation, the data from scan-in port is shifted through the scan registers to scan-out port. In the update mode, the data in scan registers is written to the optional shadow register. Scan segment might have another control port called *select* which determines whether the scan segment can perform capture, shift, and update operations.

Scan segments are connected either by buffers or *Scan Multiplexers*. The latter selects the path that scan data goes through in the network, and its select signal is referred to as *address* in the scan network literature. The internal control signals of scan segments such as *select* and the addresses of scan multiplexers are determined by the output of combinational logic blocks whose inputs are the value of shadow registers of scan segments and the primary data and control inputs of the RSN. A scan path is *active* if all the scan segments on the path are selected, and the addresses of all on-path scan multiplexers are set appropriately. To access a scan segment in the RSN, it needs to be on an active path. A read or write access to a scan segment, as defined by IEEE 1149.1 [20], is a three-step process called a *CSU* (Capture-Shift-Update) operation: in capture mode of a *CSU*, all the scan registers on the active scan path load the test result from their corresponding instrument. Then, this data will be shifted out during the followed shift operation. Note that during shift operation, the new scan data will be shifted in as the data in scan register is being shifted out. Finally, in the update mode of a CSU, the content of scan registers on the active path gets loaded to the corresponding shadow registers.

1.6.2 Segment Insertion Bit–Based RSNs

Segment Insertion Bit (SIB) is a hardware component proposed by IEEE P1687 [21] which can be used to reconfigure scan networks by bypassing or including scan chains in scan paths. In scan networks, SIBs are utilized to provide fine-grained configurable access to scan chains of instruments and their corresponding submodules. A possible implementation of the SIB is proposed in [22], which is shown in Fig. 1.8.

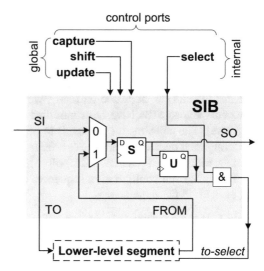

Fig. 1.8 An implementation of the segment insertion bit [22]

An SIB has a scan-input and a scan-output as well as four control inputs, capture, shift, update, and select. It also contains a 1-bit shift register S and a 1-bit shadow register U. Note that the same set of external control signals drive scan segments and SIBs in a scan network. During the shift operation, based on the value of shadow register U and the select signal, the data from scan-in port either get directed to the lower level scan segment of the SIB (*Directing mode*) or bypasses the scan segment and directly go to scan-out port (*Bypassing mode*). The value of shadow register U only gets updated from S if both *update* and *select* signals are activated. The capture operation is the same as scan segments.

The proposed IC identification scheme is based on RSNs. We will elaborate next how it can be integrated in the SIB-based RSNs; however, it could be applied on other implementations of RSN as well.

1.6.3 RSN-Based IC Fingerprinting

The proposed RSN-based IC fingerprinting scheme is built on top of the SIB-based RSNs. It takes advantage of the fact that shift register S and shadow register U in each SIB can be chained by either the Q-D or the Q'-D connection style [18–19]. In this approach, if the Q-D connection is used to chain S and U registers, the embedded ID bit is '0', and if the Q'-D connection is used, the corresponding ID bit would be '1'. Therefore, for each SIB in the design, one identification bit can be embedded.

Suppose that the original design only uses Q-D connections for all SIBs in the RSN. Then, the chip ID of this design would be all 0s. To generate a new chip ID, the designer has the option of choosing among existing SIBs to modify their

S/*U* connection styles. If *k* SIBs exist in the design, the designer can create unique digital IDs for up to 2^k chips.

As one might notice, when a Q'-D connection is used for *S*/*U* connection of an SIB, the negated value of *S* will be loaded to *U* during an update operation, which would make the original test inputs incorrect. Therefore, to ensure that all the instruments can be tested correctly, we need to adjust the test vectors for scan segments whose SIBs have been modified (Q'-D connection is used for their *S*/*U* registers). The adjustment only needs to be made to the test input, which is shifted in during each update operation. We refer to this test input as *configuration sequence* as it determines the scan network topology after its corresponding update operation.

To adjust each configuration sequence, the following rules need to be followed for each bit in the sequence.

Rule 1. If the bit corresponds to an SIB whose *S*/*U* connection style is Q'-D, the value of this bit should be set to '0' for activating the directing mode and to '1' for enabling the bypassing mode.

Rule 2. If the bit corresponds to an SIB whose *S*/*U* connection style is Q-D, the value of this bit should be set to '1' for activating the directing mode and to '0' for enabling the bypassing mode.

These rules make sure that no matter what the style of *S*/*U* connection is in each SIB, always the correct value is stored in the shadow register and scan networks can be configured correctly. In the scan network depicted in Fig. 1.9, suppose that the original design uses Q-D connections for all three SIBs, that is, the design carries an ID value of '000'. In this case, to access only scan segments 1 and 3, a configuration sequence of '101' should be shifted in before the update operation. As mentioned before, this configuration sequence only works for this specific ID, and if the S/U connection style of any SIB changes, this sequence needs to be modified. For example, if an ID equal to '101' is assigned to the scan network in Fig. 1.9, the configuration sequence for accessing scan segments 1 and 3 would be '000'.

Compared to the existing IC identification methods, our approach offers four advantages. First, it is practical as the ID bit locations in the scan network can be selected before fabrication, and the assignment of digital IDs are done in postfabrication stage. Therefore, all the designs can be fabricated with the same

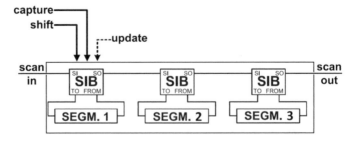

Fig. 1.9 An example of SIB-based RSN to demonstrate test input adjustments

Fig. 1.10 Programmable connections of S and U registers in ID-SIB

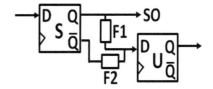

mask. Second, it incurs negligible overhead since the identification bits are added in the scan network, which won't affect the performance of core design. Third, it offers an additional nondestructive verification method which unlike other existing methods doesn't require depackaging of the IC. Finally, and most importantly, it does not require any scan chain information from each of the IP cores and is suitable for embedded devices.

To implement the presented chip identification method, we propose to replace each original SIB in the design with a slightly different version of SIB called ID-SIB. The only change we made on the original SIB is that the connection style of ID-SIB's *S* and *U* registers can be programmed in postfabrication stage, as shown in Fig. 1.10. The connection programming is done by blowing up one of the two fuses of each ID-SIB in the scan network. In Fig. 1.10, if the designer blows fuse F2, the *S*/*U* connection will be a Q-D style and the corresponding identification bit for this ID-SIB would be '0', and if she chooses to blow the other fuse, the connection would be of Q'-D style and the ID bit would be equal to '1'.

1.6.4 Security Analysis

To analyze the security of RSN-based IC fingerprinting schemes, we consider two popular attacking scenarios: fingerprinting modification and fingerprinting removal. For the proposed IC fingerprinting scheme, the removal attack can be perceived as an instance of modification attack, for removing the IC fingerprint, that is, changing all the Q'-D connections in SIBs back to Q-D connections can be viewed as a modification attack targeting IC fingerprint of all 0s. Therefore, in the following analysis, we only focus on the fingerprint modification attacks.

In fingerprint modification attacks, adversary's goal is to change the fingerprint of the chip. One possible motivation for an adversary to mount these type of attacks is to fake another chip and hopefully pass the authentication. Another one is to resell the chip to blacklisted customers for higher prices and avoid getting detected by the chip vendor. If the digital fingerprints of illegally distributed chips are not modified, the identity of the rogue customer responsible for selling these chips can be easily tracked by the chip fingerprints (or IDs).

An adversary can mount fingerprint modification attacks, only if he is capable of depackaging, reverse engineering the chip and changing the connections of *S* and *U* registers in ID-SIBs. While we believe these assumptions about capabilities of adversaries are not realistic, especially in case of very large-scale ICs, we suggest choosing fingerprint bits by the data integrity technique proposed in [17] to eliminate the possibility of such powerful attacks. Based on this technique, embedding fingerprint bits for an IC is a 4-step process: (1) choose N fingerprint bit locations and replace the corresponding SIBs with ID-SIBs, (2) choose random values for m fingerprint bits with $m < N$, (3) use this m-bit fingerprint and an IC-specific key (K_{IC}) as the input to a one-way hash function to generate (N-m) bits, (4) use the final N bits as the fingerprint bits to guide the selection of S/U connection styles at the selected fingerprint locations. In this technique, the location of the m-bit fingerprint and the value of the K_{IC} should be kept private to the IC designer or vendor.

The proposed data integrity technique makes it difficult for the attacker to forge a chip fingerprint, since a successful forgery requires knowing the value of K_{IC} and the exact location of the m-bit fingerprint, which are only known to the IC designer or vendor. Although it is possible for the adversary to change the connection styles between S and U registers in ID-SIBs, it will be challenging to make the correct changes that can maintain the property between the fingerprint bits.

1.6.5 Experimental Results

To evaluate the proposed IC fingerprinting scheme, we use the SIB-based RSN benchmarks described in [23] which are based on ITC'02 SOC benchmark set [24]. Each ITC'02 benchmark circuit is specified by the modules in the SOC and their hierarchical structure, and modules are described by the numbers of their input, output, bidirectional terminals, scan chains and their lengths, test sets, and the (x, y) coordinate of their center on the SOC layout.

In the SIB-based scan network benchmarks, two scan registers are designated for input and output pins of each module. In this design, doorway SIBs include or exclude lower-level submodules and instrument SIBs connect or bypass scan segments and input and output scan registers of each module from the active scan path, as described in [25]. In Table 1.1, the details of the ITC'02 SOC benchmarks and their corresponding SIB-based RSN designs are listed.

As described earlier, to embed the fingerprint bits, each SIB in the scan network needs to be replaced with an ID-SIB. Therefore, the number of potential fingerprint bits for each chip is equal to the number of SIBs in its scan network, which is given in Table 1.1. As one can see from the table, with the exception of q127110, all the other benchmark circuits can potentially embed a good number of fingerprint bits with the minimum of 40 bits (A586710) and maximum of 621 bits (P93791), which correspond to 1.09×10^{12} and 8.70×10^{186} unique device IDs, respectively. Even in the minimum case, 1.09×10^{12} is a couple orders of magnitude higher than the number of devices in most of the real-life embedded and IoT applications.

The proposed IC and device fingerprinting approach has negligible performance overhead as the digital fingerprint bits are only added in the SIBs of the scan

Table 1.1 Characteristics of the ITC'02 benchmarks and their corresponding SIB-based scan networks

Designs	Characteristics of the ITC'02 Benchmarks				Number of SIBs	Number of unique device IDs
	Modules	Levels	Scan segments	Register bits		
u226	10	2	40	1416	50	1.13E+15
d281	9	2	50	3813	59	5.76E+17
d695	11	2	157	8229	168	3.74E-50
h953	9	2	46	5586	55	3.60E+16
g1023	15	2	65	5306	80	1.20E+24
f2126	5	2	36	15,789	41	2.19E+12
q127110	5	2	21	26,158	25	3.35E+07
p228110	29	3	254	29,828	283	1.55E+85
p34392	20	3	103	23,119	123	1.06E+37
P93791	33	3	588	97,984	621	8.70E+186
T512505	31	2	128	76,846	160	1.46E+48
A586710	8	3	32	41,635	40	1.09E+12

network, which wouldn't cause any overhead to the design. Moreover, the overhead incurred on testing instruments is also negligible since no extra hardware is integrated into the design, and all the changes are local, which avoids rerouting. For different RSN configurations, the testing vector can be justified, which is a one-time cost, so there will no change in test coverage.

1.7   Summary

Most of the Internet of Things (IoT) and embedded devices are resource constrained, making it impractical to secure them with the traditional computationally expensive crypto-based solutions. However, security and privacy are crucial in many IoT applications. The embedded devices, or the things in IoT, are the root for IoT, and their security and trust are also the foundation for IoT security. This book is dedicated to device authentication. In this chapter, we give the motivation for this work and argue that lightweight authentication schemes based on hardware features are promising for IoT applications. One recent popular research area is IC fingerprint-based device authentication. As the foundation for the book, we discuss the general requirements for any effective device authentication based on fingerprint, the specific ones for IoT applications, the standard 2-phase authentication protocol, as well as fingerprints in analog and digital. Then we focus on the existing IC fingerprinting techniques and explain the core idea, basic procedure, advantages, and limitations for each of these techniques. Finally, we propose a practical approach using the reconfigurable scan network (RSN), which is a standard testing feature for IC design.

Acknowledgments This work is supported by the National Natural Science Foundation of China under Grant No. CNS1745466 and Air Force Office of Scientific Research MURI under award number FA9550-14-1-0351. The authors would also want to thank their colleagues Mr. Omid Aramoon, Dr. Aijiao Cui, and Dr. Carson Dunbar.

References

1. G. Qu, L. Yuan, Design Things for the Internet of Things – An EDA Perspective, *IEEE/ACM International Conference on Computer Aided Design (ICCAD'14)*, pp. 411–416, November 2014
2. J. Steinberg, These Devices May Be Spying On You (Even In Your Own Home). *Forbes*. 27 January 2014
3. C. Dunbar, G. Qu, A DTN Routing Protocol for Vehicle Location Information Protection, *Military Communications Conference (Milcom'14)*, October 2014
4. J. Lach, W. H. Mangione-Smith and M. Potkonjak, FPGA Fingerprinti-ng Techniques for Protecting Intellectual Property, *Proceedings of CI-CC*, 1998
5. A.E. Caldwell, H. Choi, A.B. Kahng, et al, Effective Iterative Techniques for Fingerprinting Design IP, *36th ACM/IEEE Design Automation Conference (DAC'99)*, pp. 843–848, June 1999
6. A.E. Caldwell, H. Choi, A.B. Kahng, et al., Effective iterative techniques for fingerprinting design IP. IEEE Transactions on Computer-Aided Design of Integrated Circuits and Systems **23**(2), 208–215 (February 2004)
7. G. Qu, M. Potkonjak, Fingerprinting Intellectual Property Using Constraint-Addition, *37th ACM/IEEE Design Automation Conference (DAC'00)*, pp. 587–592, June 2000
8. G. Qu, M. Potkonjak, *Intellectual Property Protection in VLSI Designs: Theory and Practice*, Kluwer Academic Publishers, ISBN 1-4020-7320-8, January 2003
9. K. Machida, S. Shigematsu, H. Morimura, et al., A novel semiconductor capacitive sensor for a single-Chip fingerprint sensor/identifier LSI. IEEE Transactions on Electron Devices **48**(10), 2273–2278 (October 2001)
10. F. Koushanfar, G. Qu, M. Potkonjak, Intellectual Property Metering, *4th Information Hiding Workshop (IHW'01)*, pp. 87–102, LNCS Vol. 2137, Springer-Verlag, April 2001
11. F. Koushanfar, G. Qu, Hardware Metering, *38th ACM/IEEE Design Automation Conference (DAC'01)*, pp. 490–493, June 2001
12. F. Koushanfar, Hardware metering: A survey, in *Introduction to Hardware Security and Trust*, ed. by M. Tehranipoor, C. Wang, (Springer, 2012)
13. C. Dunbar, G. Qu, Satisfiability don't care condition based circuit fingerprinting techniques, *20th Asia and South Pacific Design Automation Conference (ASPDAC'15)*, pp. 815–820, January 2015
14. C. Dunbar, G. Qu, A practical circuit fingerprinting method utilizing observability don't care conditions, *Design Automation Conference (DAC'15)*, June 2015
15. G. Qu, C. Dunbar, X. Chen, A. Cui, Digital fingerprint: A practical hardware security primitive, in *Digital Fingerprinting*, pp. 89–114, Springer, ISBN 978-1-4939-6599-1, 2016
16. N.R. Wagner. Fingerprinting. *Proceedings of the 1983 Symposium on Security and Privacy, IEEE Computer Society*, pp. 18–22, 1983
17. X. Chen, G. Qu, A. Cui, C. Dunbar, Scan chain based IP fingerprint and identification. *18th International Symposium on Quality Electronic Design (ISQED)*, 2017
18. A. Cui et al., Utral-low overhead dynamic watermarking on scan Design for Hard IP protection. IEEE Transactions on Information Forensics and Security **10**, 2298–2313 (July 2015)
19. S. Gupta, T. Vaish, S. Chattopadhyay, Flip-flop chaining architecture for power-effcient scan during test application. Proceedings of Asia Test Symposium, 410–413 (December 2005)
20. IEEE standard for test access port and boundary-scan architecture, IEEE Standard 1149.1-2013, 2013

21. IJTAG, "IJTAG - IEEE P1687," March 2012. [Online]. Available: http://grouper.ieee.org/groups/1687
22. R. Baranowski, M.A. Kochte, H.J. Wunderlich, Fine-grained access Management in Reconfigurable Scan Networks. IEEE Transactions on Computer-Aided Design of Integrated Circuits and Systems **34**(6), 937–946 (June 2015)
23. R. Baranowski, M. A. Kochte, H. J. Wunderlich, Modeling, verification and pattern generation for reconfigurable scan networks, *IEEE International Test Conference*, Anaheim, CA, 2012, pp. 1–9
24. E. J. Marinissen, V. Iyengar, K. Chakrabarty, A set of benchmarks for modular testing of SOCs, *Proceedings. International Test Conference*, 2002, pp. 519–528
25. F. G. Zadegan, U. Ingelsson, G. Carlsson, E. Larsson, Design automation for IEEE P1687, *Design, Automation & Test in Europe*, Grenoble, 2011, pp. 1–6

Chapter 2
Physical Unclonable Function: A Hardware Fingerprinting Solution

Mohd Syafiq Mispan and Basel Halak

Abstract Physically unclonable functions or PUFs are innovative hardware security primitives which produce unclonable and inherent device-specific identifier of particular hardware. The notion of PUFs is a resemblance to the biometric fingerprint of human beings. The inherent device-specific identifier is produced by exploiting the intrinsic process variations during integrated circuit (IC) fabrication. IC fabrication imposes variability in oxide thickness, dopant implantation, line-edge roughness, etc. which has an impact on the electrical behavior of MOSFET transistors. These electrical mismatches are extracted and manifested as random and unique PUF responses. Since PUFs extract unique hardware characteristics, they potentially offer an affordable and practical solution in the hardware-assisted security field. In this chapter, we present the concept, properties, and construction of PUFs which make them a promising solution for identification and authentication, and secret key generation applications.

Keywords PUF · Process variability · Integrated Circuits (IC) ·
Nanotechnology · FinFET · CMOS · Memristor · IC fingerprinting ·
Cryptographic key generation

M. S. Mispan
Universiti Teknikal Malaysia Melaka, Fakulti Teknologi Kejuruteraan Elektrik & Elektronik, Melaka, Malaysia

Centre for Telecommunication Research & Innovation (CeTRI), Micro & Nano Electronics, Malaysia
e-mail: syafiq.mispan@utem.edu.my

B. Halak (✉)
University of Southampton, Southampton, UK
e-mail: Basel.Halak@soton.ac.uk; mz@soton.ac.uk

© Springer Nature Switzerland AG 2021
B. Halak (ed.), *Authentication of Embedded Devices*,
https://doi.org/10.1007/978-3-030-60769-2_2

2.1 Introduction

The inception of a network of smart objects ("things") or the so-called Internet of Things (IoT) allows these smart electronic devices to be interconnected, uniquely identifiable, and controlled remotely through the Internet technology. The interconnection enables collecting, transferring, processing, and analyzing a huge amount of data within a set of connected devices. IoT technology is expected to improve human productivity and quality of life shortly soon. The IoT can be applied in various fields such as system automation, medical and health-care, remote sensing, secure access, agriculture, etc. This wide range of applications process sensitive and user-specific data which requires robust hardware and strict security protocol to avoid potential security breaching that could lead to loss of privacy. With the incoming and abundance of the IoT devices, a secure, reliable and trustworthy root-of-trust for identification, authentication, and integrity checking of the system are becoming paramount importance for further development of the IoT technology.

Implementation of current security solutions relies on the secret keys stored in the on-chip non-volatile memory (NVM) or battery-backed static random-access memory (SRAM) [5, 20]. These security methods based on secret digital keys often do not provide adequate solutions for the aforementioned purposes. The secret keys are not inherent (i.e., programmed by the system owner) and the keys are always present in the system. Hence, the secret keys stored in NVM are vulnerable to the extracting and cloning type of attacks [37]. Moreover, the cost of implementation of the NVM or battery-backed SRAM key storage solutions is considered high for resource-constrained pervasive devices.

In recent years, *Physical Unclonable Functions* (PUFs) have emerged as a promising solution to provide a hardware root-of-trust for integrated circuit (IC) applications and can answer the critical security-related issues as discussed above with a relatively low cost. PUF operating principle is based on the exploitation of nano-scale device-level intrinsic process variations from which device-specific random keys are derived [30]. In IC applications, the intrinsic process variations are undesired effect, generally for most of the circuitries, but it is essential for a functional PUF. A unique response extracted from a PUF acts as an electronic fingerprint that uniquely distinguishes each PUF from a group of similar PUFs. Therefore, there is no need to store a secret key in any memory devices, unlike the conventional security solutions as discussed above. PUFs are seen to have some advantages such as energy-efficient, low-overhead, low manufacturing cost, non-programmable, and easy to manage secret keys.

As PUFs extract unique hardware characteristics, it can be used to uniquely identify a hardware and also can be used to bind the firmware or software, hence providing trustworthy to the whole system. Therefore, in this chapter, we will explore the details of the PUF concept and properties. Some of the PUF architectures will be discussed to understand the impact of process variations and the manifestation of their unique and random responses. PUF evaluation metrics and its application as hardware-assisted security solutions will be discussed as well.

2.2 Chapter Overview

This chapter is organized as follows: Sect. 2.3 describes the details of the PUF concept and properties which resemblance to human biometric fingerprint. This section also describes the process variability in the IC and the history of silicon PUFs. Section 2.4 provides an overview of the development in PUF constructions. The PUF evaluation metrics are also discussed in this section which are used to quantify the performance of a particular PUF. Section 2.5 describes the application of PUFs as authentication and identification devices and a cryptographic key generator. Finally, Sect. 2.6 provides the conclusion for this chapter.

2.3 Physical Unclonable Function

2.3.1 PUF Concept and Properties

The basic concept of PUF is an exploitation of a function embodied in a physical material of the device. This function is non-deterministic and varies from one instance to another of the same material [27]. In this chapter, the focus is on silicon PUFs, which exploit the intrinsic and random variations in CMOS devices during IC fabrication. By exploiting the IC manufacturing process variations, PUFs can map a set of challenges to a set of random responses, known as challenge-response pairs (CRPs). As illustrated in Fig. 2.1, two PUF instances of similar types of PUF (PUF A and PUF B) produced two unique responses where Response A $\neq$ Response B when the same challenge is applied. Therefore, the output of the PUF is random and device-specific which has a potential to be used as hardware fingerprinting to improve the security for any given applications.

According to [8], PUFs should have the two properties which are easy to evaluate and hard to characterize. Elsewhere, [15] interpreted PUFs as a complex uncontrollable system that have the following properties:

1. The set of CRPs (C_i, R_i), $i = 1, \ldots N$ is reproducible and can be repeatedly extracted.
2. Given a set of CRPs, (C_i, R_i), it is impossible to build the mathematical model to compute and simulate another set of CRPs, (C_j, R_j), where $i \neq j$, of similar type of PUF.
3. It is impossible to physically reproduce or cloning as an analogous system having an identical set of CRPs, (C_i, R_i).

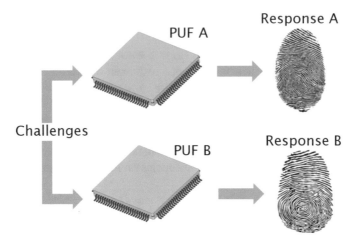

Fig. 2.1 Basic functionality of PUF

Fig. 2.2 LER and RDF,
variation sources in a
nano-scale device

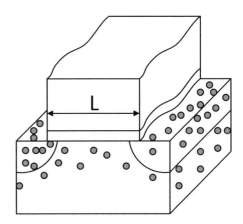

2.3.2 Process Variability in Integrated Circuits

The aggressive scaling of CMOS technology has led to a drastic increase in process
variations such as oxide thickness, random dopant fluctuations (RDF), line-edge
roughness (LER), and other layout-dependent effects which causes a direct impact
on the electrical behavior of MOSFETs. According to Ye et al. [44], LER and
RDF are primary intrinsic variation sources in CMOS structure. Both effects are
illustrated in Fig. 2.2.

RDF is described as the fluctuations of the total number of dopants in the
transistor channel which vary randomly from transistor to transistor [22]. This
fluctuation is mainly affected by the randomness in the amount and position of
dopants during dopant implantation. As devices scale down, the total number of

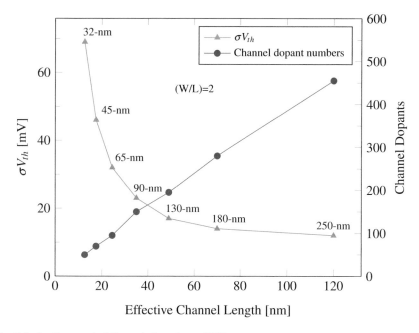

Fig. 2.3 Scaling trend of V_{th} variations due to RDF

channel dopants decreases due to the channel volume decrease. As a result, the relative effect of a single change in dopant number increases and the variations in the V_{th} become significant [44]. Figure 2.3 illustrates the effect of RDF on different technologies ranging from 250-nm to 32-nm technology node. The variation in the V_{th} of MOSFETs which is represented as σV_{th} is significantly determined by the number of dopants in the transistor channel, as can be seen in Fig. 2.3.

On the other hand, LER is described as the distortion of gate shape along channel width direction as indicated in Fig. 2.2. The variations due to LER are inherent to gate materials and are mainly affected by the process of gate etching. Due to these facts, its variance does not scale with the technology which raises a major concern [1]. Furthermore, the improvement in the lithography process does not effectively reduce its variations [1]. Figure 2.4 depicts the impact of LER on V_{th} variations which was numerically simulated using a 65-nm predictive technology model (PTM) [48]. The relationship between V_{th} and transistor width in Fig. 2.4 shows that the impact of LER on V_{th} variations is profound when the transistor width is small. LER interacts with RDF, profoundly impacting all aspects of circuit performance, especially in the design of CMOS circuits which are extremely sensitive to mismatch. Although process variation is an undesired effect for most of the CMOS circuitry, it is the desired and essential effect for PUFs.

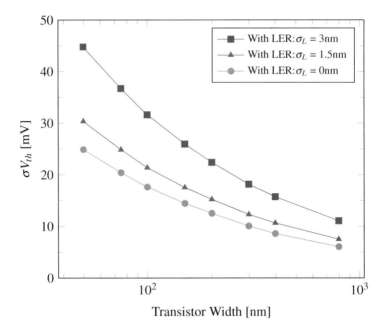

Fig. 2.4 Impact of LER on V_{th} variations using 65-nm PTM

2.3.3 Chronology of Silicon Based PUFs

The inception of integrated circuit identification (ICID) using device mismatch was proposed by Lofstrom et al. [22]. The main motivation is to reduce the manufacturing cost and to produce a non-programmable identifier or key. ICID exploits the randomness inherent in silicon processing to extract unique and repeatable information. ICID was fabricated using a 0.35 μm single-poly n-well process. Nevertheless, the name of PUF or any part of it is not mentioned in [22], but the concept was very similar to that of a PUF. Subsequently, physical one-way functions (POWFs) are proposed by [34]. The POWF is based on the optical principle of laser lights which applied on scattering particles in a transparent optical medium. The laser lights were applied at a different angle, distance, and wavelength resulted in the speckle patterns which are found to be unique and unpredictable. The concept of POWF has led to the idea of PUF.

The notion and the name of "Physical Unclonable Function" were first revealed in [8]. The idea of PUF which exploits the intrinsic variations during IC fabrication is elaborated and the preliminary experiments are conducted using Xilinx XC2S200 Field Programmable Gate Arrays (FPGAs). The first silicon implementation of PUF has been presented in [19]. Lee et al. [19] proposed an Arbiter-PUF architecture and fabricated on silicon using TSMC 180-nm technology. Thereafter, many different types of PUFs have been proposed such as ring oscillator PUF (RO-PUF), [38], static random-access memory PUF (SRAM-PUF), [9, 13], Butterfly PUF, [18], etc.

2.4 PUF Constructions

Generally, PUF can be divided into two classes which are Strong PUFs and Weak PUFs. This classification has been first introduced in [9] and later developed further by Rührmair et al. [35]. Strong PUFs are PUFs with an exponential number of CRPs, 2^i where the CRPs is given as (C_i, R_i), $i = 1, \ldots N$. Weak PUFs are PUFs with a very small number of CRPs, fixed challenges, and in the extreme case with just only a single challenge [35]. The "Strong" and "Weak" terms are not meant to indicate that one PUF-type would be superior or inferior to another [35]. In this section, however, we give an overview of PUF based on the development and evolution from the architecture and technology standpoint.

2.4.1 Delay, Mixed-Signal, and Memory PUFs

Since the inception of PUF [8], the development and exploration in PUFs construction are progressed drastically. At the early development of PUF, generally, PUF architectures can be categorized into delay, mixed-signal, and memory-based PUFs. One of the delay based PUFs is known as Arbiter-PUF. Arbiter-PUF is the earliest PUF which have been proposed and fabricated on silicon using TSMC 180-nm technology node [19]. Figure 2.5 shows the design of k-bit Arbiter-PUF which consists of k switching components and an Arbiter. The switching component is built using a 2-to-1 multiplexer. A rising pulse at the input propagates through the switching components. k-bit challenge, $C = (c_1, c_2, \ldots, c_k)$ is used to control the propagation path of the rising pulse applied at the input. For $c_k = 0$, the paths go straight, while for $c_k = 1$, they are crossed. As explained in Sect. 2.3.2, because of intrinsic process variations in each switching component, there is a delay difference of Δt between the paths. Subsequently, an Arbiter which constructed based on SR-latch or D flip-flop is triggered by the delay difference and a response, r is generated. Since the process variations affect the delay difference of k switching components, the generation of response r will be random and device-specific which has the potential to be used as a secret key or an identifier. Total number of CRPs that can be generated by k-bit Arbiter-PUF is 2^k. Hence, Arbiter-PUF can be classified as Strong PUF.

The introduction of Arbiter-PUF naturally led to other PUF variants which derived from Arbiter-PUF architecture such as XOR Arbiter-PUFs, Lightweight-PUFs, and Feedforward Arbiter-PUFs [21, 25, 38]. These PUFs are proposed to inject non-linearity into the PUF architecture (Fig. 2.6). The non-linearity helps to increase the mapping complexity of the CRPs, hence achieving the ideal PUF properties as described earlier in Sect. 2.3.1.

In a study, Cao et al. [3] proposed a mixed-signal PUF that based on a widely used CMOS image sensor circuit. The function of the image sensor is not affected when operated in a PUF mode. The CMOS image sensor consists of an array of

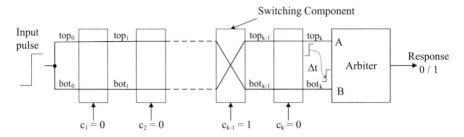

Fig. 2.5 k-bit Arbiter-PUF design

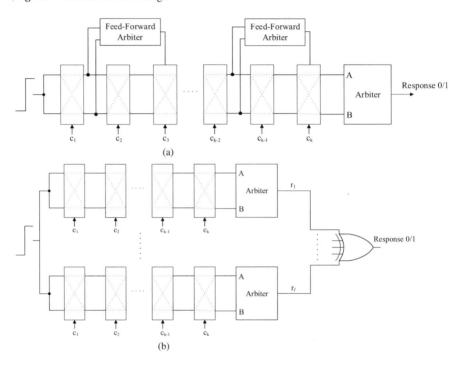

Fig. 2.6 Arbiter-PUF derivatives. (**a**) k-bit Feedforward Arbiter-PUFs. (**b**) l XOR k-bit Arbiter-PUFs

pixel circuits. The typical structure of a pixel array circuit is illustrated in Fig. 2.7. During a reset phase, M_{RS} is turned on and the voltage on the photodiode (PD), V_{PD} is given as:

$$V_{PD} = V_{dd} - V_{th,RS} \qquad (2.1)$$

where $V_{th,RS}$ is the threshold voltage of M_{RS} and V_{dd} is the power supply. When M_{RS} is turned off, the PD is exposed to the illumination for exposure time, t. After t, M_{SEL} is turned on and the output voltage, V_{out} of the pixel cell is read out, given as:

Fig. 2.7 3-T pixel circuit

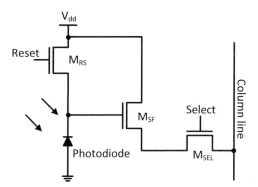

$$V_{out} = V_{dd} - V_{th,RS} - V_{th,SF} - \frac{I_{PH} \times t}{C_{PD}} \quad\quad (2.2)$$

where $V_{th,SF}$ is the threshold voltage of M_{SF}, I_{PH} is the photocurrent, and C_{PD} is the PD junction capacitance. This PUF exploits the variations in the pixel output voltage which is mainly caused by the variations in the $V_{th,SF}$, $V_{th,RS}$, and C_{PD}. A random response, "0" or "1" is generated by comparing two output voltages of two pixels, based on which reset voltage is larger.

Elsewhere, a memory-based PUF using SRAM circuit has been proposed by Guajardo et al. [9] and Holcomb et al. [13], known as SRAM-PUF. The random response of SRAM-PUF is generated after the power-up process. Figure 2.8 depicts the 6-T SRAM cell circuit which consists of a back-to-back inverter (MP1, MN1, MP2, and MN2) and two access transistors (MN3 and MN4). When an SRAM cell is powering-up, as the supply voltage increases, the current flowing through MN1 and MN2 will slowly pull up the voltage at nodes Q and QB. Due to the intrinsic process variations, the V_{th} of transistor MP1 is slightly higher compared to that of MP2. Therefore, the current that flows through MN2 is slightly higher than through MN1, thus turning ON the MN1 and pulling down node Q to GND. At the same time when node Q is discharging, MP2 is turned ON and pulls up node QB to V_{dd}. The internal nodes of bi-stable SRAM, Q, and QB settle at "0" and "1," respectively, as illustrated in Fig. 2.9. When powering-up the SRAM, the start-up values (SUVs) across different memory blocks within an SRAM and across multiple SRAMs show random patterns and are device specific, which are the desired qualities to be used as a PUF. An SRAM-PUF is considered as a Weak PUF since it has only a single challenge which is a power-up process to generate random SUVs. One might argue that addressing the SRAM bit cell array provides a challenge-response mechanism, however, it is only the process of reading the bit cell values.

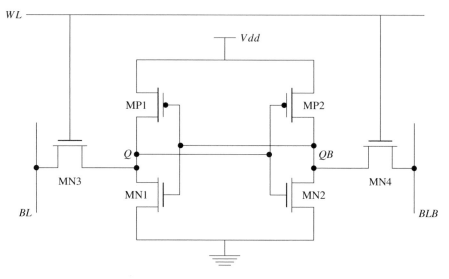

Fig. 2.8 6-T SRAM cell circuit

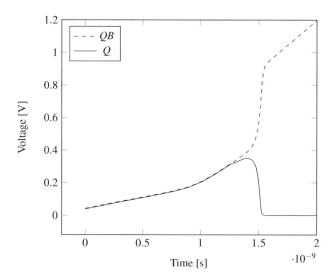

Fig. 2.9 Internal nodes of SRAM cell, Q and QB resolving to 1 and 0 due to V_{th} variation during the power-up process

2.4.2 FinFET Based PUFs

The PUFs circuitry proposed earlier in the literature including PUFs which have been discussed in Sect. 2.4.1 is constructed based on a planar metal-oxide-semiconductor field-effect transistor (MOSFET). As CMOS technology scales

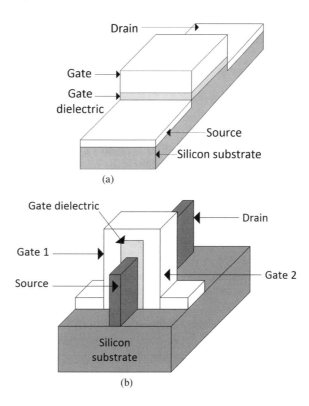

Fig. 2.10 Structural comparison between (**a**) planar bulk MOSFET and (**b**) bulk FinFET

down, the short-channel effects are significantly degrading the performance of the MOSFET transistors. To further continue the scaling down of CMOS technology, FinFET is introduced to replace 2D (planar) MOSFET. Figure 2.10 illustrates the structural comparison of planar bulk MOSFET and FinFET [2]. The feasibility of FinFET has been proven as Intel has manufactured its Ivy Bridge processor in 2011 using 22-nm FinFET [14].

In the PUF research area, FinFET is only recently being explored. Zhang et al. [47] studied the feasibility of constructing SRAM-PUF using FinFET technology which expected to have advantages of small feature size and better randomness due to increase in FinFET's variability. LER has been assumed as a source of process variations that affect the fluctuation in V_{th}. LER is modeled using a Gaussian autocorrelation function. The findings in [47] show that the FinFET based SRAM-PUF is feasible with good randomness and the SUVs reliability of 88% (i.e., 12% bit error due to CMOS noise).

Elsewhere, Narasimham et al. [32] analyzed the SRAM-PUF quality and reliability built using 28-nm high-k metal gate (HKMG) planar MOSFET and 16-nm FinFET processes. The quality analysis indicates that FinFET shows better randomness over planar MOSFET as the number of bit cells that power-up to "1" and "0" is balanced. The reliability of both planar and FinFET has been evaluated under aging through burn-in stress at high temperature, 125°C, and high supply voltage,

Fig. 2.11 TiO$_2$ based
memristor structure

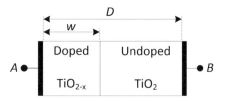

1.4× nominal. The reliability analysis indicates that the FinFET is more susceptible to aging and has a marginally higher bit error rate compared to that HKMG planar MOSFET.

2.4.3 Nanotechnology PUFs

Earlier, most PUF designs focus on exploiting process variations intrinsic of the CMOS technology. In recent years, progress in emerging nanoelectronic devices has demonstrated a more severe level of intrinsic process variations as a consequence of scaling down to the nano region [7]. In a study, [26] a novel hybrid memristor-CMOS based PUF is proposed which provides excellent statistical randomness and consumes small area overhead. The memristor is a non-volatile electronic memory device. Figure 2.11 depicts the basic structure of a memristor which composed of a highly doped TiO$_{2\text{-}x}$ layer with oxygen vacancy and an almost undoped TiO$_2$ without oxygen vacancy placed between a pair of electrodes A and B. To program the memristor, a voltage is applied across the electrodes, oxygen atoms in the material diffuse right or left, depending on the voltage polarity, which makes the $w(t)$ thinner or thicker, thus producing a change in resistance. Once the voltage is removed, the resistance value is retained. The equivalent resistance of the memristor device is given as [26]:

$$R_{eq}(t) = \frac{w(t)}{D}.R_{on} + \left(1 - \frac{w(t)}{D}\right)$$
(2.3)

where R_{on} is the low resistance of the device if the entire device is doped, and R_{off} is the high resistance of the device if the entire device is undoped. The impact of process variations on the thickness parameter, D is highly nonlinear and extremely pronounced, resulting in the random values of memristors resistance which can differ by several orders of magnitude.

The proposed PUF, [26] inherits the structure and the delay-variations effects of an Arbiter-PUF (see Fig. 2.5). The architecture of the proposed memristor-PUF is illustrated in Fig. 2.12 which consists of the timing-and-control unit, two identically laid out memristor delay paths, reset elements, and nMOS switches between two consecutive memristors. During the reset phase ($V_{RST} = 1$), an effective potential difference across each memristor device is built up differently

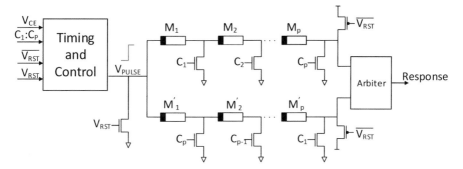

Fig. 2.12 Architecture of the memristor-CMOS based PUF

and memristors are in a random resistance due to intrinsic process variations. When $V_{RST} = 0$, the challenge enable voltage, V_{CE} and the pulse voltage, V_{pulse} are applied simultaneously. After a delay, t_{CE}, the V_{CE} is set to "0" and the V_{pulse} is propagated to the end of the delay paths. The arbiter generates a random response of "0" or "1" depending on which of the two pulses reach the arbiter first.

Elsewhere, Gao et al. [6] proposed a memristor-based nanocrossbar PUF architecture, known as mrPUF. A crossbar architecture is an array of crossing nanowires and the memristor is formed at the junction between two crossing wires. Figure 2.13 illustrates memristor-based PUF with a nano crossbar array. The red marked memristors are the selected memristors when the challenge is applied. Both selected memristors are used to control the current in the current mirror-controlled ring oscillator (CM-RO). The CM-RO manifests the variations in memristors into the frequency and subsequently, digitized by the counter and comparator blocks as a random binary response, "0" or "1."

The memristor-based PUF discussed above has a unique property which is cycle-to-cycle (C2C) variations introduced by each programming operation. Hence, a memristor could be exploited to build a reconfigurable PUF. Other nanotechnologies such as carbon-nanotube field-effect transistor (CNFET) based PUF, [17] and phase-change memory (PCM) based PUF [46] are also discussed in the literature. Nanotechnology-based PUFs offer a few advantages over CMOS based PUFs, such as substantial process variations, small foot-prints, and lower energy consumption [7].

2.4.4 PUF Evaluation Metrics

All of the above described an overview of PUF development in general. In actual fact, enormous PUFs architectures have been proposed in the literature. Therefore, a standard set of parameters is needed to quantify the performance of the proposed

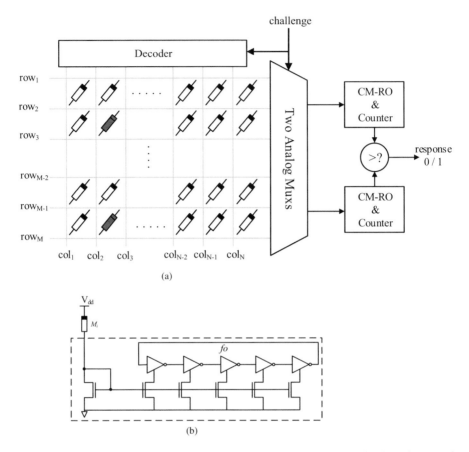

Fig. 2.13 Memristor-based nanocrossbar PUF architecture, mrPUF. (**a**) Top level architecture of mrPUF. (**b**) Current mirror-controlled ring oscillator (CM-RO)

PUFs. Maiti et al. [24] have systematically defined the quality parameters to evaluate and compare the performance of PUFs which will be discussed in this section.

2.4.4.1 Uniqueness

The response generated from a PUF is evaluated based on its uniqueness. A uniqueness measures the probability of one PUF instance being uniquely distinguished from a group of PUFs of a similar type and it has an ideal probability value of 50%. A probability of 50% means for the same challenge applied to two similar PUFs, one PUF will generate a response of 50% different compared to the other PUF. The hamming distance (HD) is used to evaluate the uniqueness performance and is called the "Inter-HD." The HD between two equal-length binary strings is the number of positions at which corresponding bits are different. If the same challenge

C is applied to two chips, i and j ($i \neq j$), and n-bit responses are generated, $R_i(n)$ and $R_j(n)$, respectively, the average Inter-HD among k chips is defined as [24]:

$$\text{Inter–HD} = \frac{2}{k(k-1)} \sum_{i=1}^{k-1} \sum_{j=i+1}^{k} \frac{HD(R_i(n), R_j(n))}{n} \times 100\% \qquad (2.4)$$

2.4.4.2 Reliability

The reliability is a measure of the reproducibility of the PUF responses given the same challenge under volatile conditions of ambient temperatures and/or supply voltage. Similarly to uniqueness, the HD is used to evaluate the reliability and is called the "Intra-HD." To estimate the Intra-HD, an n-bit reference response $R_i(n)$ is extracted at room temperature and a nominal supply voltage (i.e., the reference condition) given a challenge C, for a single chip, represented as i. The same challenge C is applied to the chip i at different condition to obtain the n-bit response, $R'_{i,j}(n)$. Hence, the average Intra-HD for m samples is defined as, [24]:

$$\text{Intra–HD} = \frac{1}{m} \sum_{j=1}^{m} \frac{HD(R_i(n), R'_{i,j}(n))}{n} \times 100\% \qquad (2.5)$$

From the Intra-HD value, the reliability of a PUF can be defined as:

$$\text{Reliability} = 100\% - \text{Intra–HD} \qquad (2.6)$$

From Eq. (2.6), a small Intra-HD is desired to achieve high reliability. Aside from temperature and supply voltage fluctuations, previous literature [23, 28, 31, 32] reported that device aging processes such as bias temperature instability (BTI), hot-carrier injection (HCI), and temperature-dependent dielectric breakdown (TDDB) can cause bit errors in PUF response. Device aging manifest itself as an increase in threshold voltage, V_{th}. According to [28], Eq. (2.5) and (2.6) can be extended to compute bit errors due to device aging effect. If a single chip, represented by i, responds to the challenge C with an n-bit reference response $R_{i.fresh}(n)$ at $t = 0$ and an n-bit response, $R_{i.aged}(n)$ at $t = t$, the average Intra-HD$_{\text{aging}}$ for m samples is defined as:

$$\text{Intra–HD}_{\text{aging}} = \frac{1}{m} \sum_{i=1}^{m} \frac{HD(R_{i.fresh}(n), R_{i.aged}(n))}{n} \times 100\% \qquad (2.7)$$

From the Intra-HD$_{\text{aging}}$ value, the reliability of a PUF due to device aging effect can be defined as:

$$\text{Reliability}_{\text{aging}} = 100\% - \text{Intra–HD}_{\text{aging}} \qquad (2.8)$$

2.4.4.3 Uniformity

The uniformity is defined as the proportion of 0s and 1s in the response bits of a PUF which characterize the randomness of the PUF response. The proportion of 0s and 1s must be balanced, hence ideally the value of uniformity is distributed at 50%. The hamming weight (HW) is used to evaluate the uniformity which measures number "1" bits in the binary sequence, and is described as below, [24]:

$$\text{Uniformity} = \frac{1}{k \times n} \sum_{i=1}^{k} \sum_{j=1}^{n} r_{i,j} \times 100\% \tag{2.9}$$

where $r_{i,j}$ is the j-th binary bit of an n-bit response from a chip i, for a total of k chips.

Uniformity and uniqueness quality metrics as discussed above are independent parameters. For k chips of a similar type of PUF with an n-bit response from each chip, the average uniformity, can be close to an ideal value of 50% but that does not guarantee 50% uniqueness. For example, k chips of the worst PUF could generate k similar n-bit responses which have a balanced distribution of 0s and 1s in their n-bit responses. On the other hand, k chips of a similar type of PUF with an n-bit response from each chip can achieve a uniqueness close to an ideal value of 50% but the average uniformity is not necessarily at 50%. For example, one or more of the k chips of the worst PUF could generate all 1s or 0s in their corresponding n-bit responses.

2.5 PUFs Applications

Two categories of applications naturally emerge, linked to functional discrepancies of Strong and Weak PUFs, which are IC identification and authentication, and cryptographic key generation, respectively [5, 11]. Nevertheless, the use of a certain type of PUFs is not, however, limited to a certain application as described in [10, 40]. Therefore, generally, all types of PUFs can be used in any security-related applications which require a unique identifier or secret key. In this section, a generic framework of PUF-based identification and authentication, and cryptographic key generation applications are discussed.

2.5.1 IC Identification and Authentication

The CRPs of PUF can be used in IC identification and authentication as they can be used as the basis of challenge-and-response protocol. Figure 2.14 depicts a basic of PUF-based authentication using the challenge-and-response protocol. The protocol

Verifier $\longrightarrow$ **Prover** j

$\langle \mathbf{c}_{ij}, \mathbf{r}_{ij} \rangle$ with $\mathbf{c}_{ij} \leftarrow TRNG()$ and $i \in [1\ d]$ $\longleftrightarrow$ $\mathbf{r}_{ij} \leftarrow PUF(\mathbf{c}_{ij})$ $\Big\}$ 1x Enrolment

$d_j \leftarrow d$

$\langle \mathbf{c}, \mathbf{r} \rangle \leftarrow \langle \mathbf{c}_{ij}, \mathbf{r}_{ij} \rangle$ with $i \leftarrow d_j$

$d_j \leftarrow d_j - 1$ $\xrightarrow{\ \mathbf{c}\ }$ $\hat{\mathbf{r}} \leftarrow PUF(\mathbf{c})$ $\Big\}$ dx Authentication

Abort if $\mathbf{HD}(\mathbf{r}, \hat{\mathbf{r}}) > \varepsilon$ $\xleftarrow{\ \hat{\mathbf{r}}\ }$

Fig. 2.14 Basic PUF-based authentication protocol

consists of two phases; (1) the enrolment phase and (2) the authentication phase. The enrolment phase is performed once in a controlled and trusted environment. During this phase, the verifier sends a set of randomly chosen challenges to obtain unpredictable responses from the authentic device, prover j. These CRPs are stored in a database, d_j for future authentication.

Once the prover j is deployed in the field and it is requested for authentication, the verifier selects a challenge from d_j, send to the prover j and obtains the response $\hat{r}$. The response $\hat{r}$ is compared against the stored response in a database. If the hamming distance (HD) is less than the threshold mismatch, ϵ, the prover j is identified as an authentic device.

The basic authentication protocol as described in Fig. 2.14 is susceptible to model building attack by using machine learning (ML) techniques [29, 42]. Hence, it leads to further development of PUF-based authentication protocols to enhance its security against ML-attack. A comprehensive survey of several techniques on PUF-based lightweight authentication protocols is well described in [5]. One of the methods to improve the security against ML-attack is by using hash functions as depicted in Fig. 2.15. Similarly, the protocol consists of two phases which are enrolment and authentication phases. *Gen* is executed only once during the enrolment phase to generate a helper data, **p**. The helper data **p** is stored by the verifier or in NVM of the prover. *Rep* is executed at each of the authentication processes. The helper data **p** is recalled during the execution of *Rep* to reproduce the exact PUF response as previously generated during the enrolment phase, before generating the final response, $\hat{r}$. If the response of the prover j, $\hat{r}$ is matched with the response in the verifier's database, r, the prover j is identified as an authentic device.

The above PUF-based authentication protocol has the potential to be promoted as a secure alternative to memory-based radio frequency identification (RFID) tags. A current identification and authentication practice of memory tags RFID is by storing unique identifiers in NVM. As mentioned earlier in Sect. 2.1, the unique identifier or secret key which always present in the NVM is vulnerable to the extracting and cloning type of attacks [37]. Therefore, the above protocol is suitable to be applied for PUF-based RFID. The development of PUF-based

Verifier $\longrightarrow$ **Prover** j

$\langle \mathbf{c}_{ij}, \mathbf{r}_{ij}, \mathbf{p}_{ij}\rangle$ with $\mathbf{c}_{ij} \leftarrow TRNG()$ $\longleftarrow$ $\mathbf{r}_{ij} \leftarrow Hash(PUF(Hash(\mathbf{c}_{ij})), Hash(\mathbf{c}_{ij}))$ $\left.\right\}$ 1x Enrolment
and $i \in [1\ d]$ $\mathbf{p}_{ij} \leftarrow Gen(PUF(Hash(\mathbf{c}_{ij})))$
$d_j \leftarrow d$

$\langle \mathbf{c}, \mathbf{r}, \mathbf{p}\rangle \leftarrow \langle \mathbf{c}_{ij}, \mathbf{r}_{ij}, \mathbf{p}_{ij}\rangle$ with $i \leftarrow d_j$

$d_j \leftarrow d_j - 1$ $\xrightarrow{\ \mathbf{c}\ }$ $\hat{\mathbf{r}} \leftarrow Hash(Rep(PUF(Hash(\mathbf{c})), \mathbf{p}), Hash(\mathbf{c}))$ $\left.\right\}$ dx Authentication

Abort if $\mathbf{r} \neq \hat{\mathbf{r}}$ $\xleftarrow{\ \hat{\mathbf{r}}\ }$

Fig. 2.15 PUF-based authentication protocol with hash functions

authentication protocol, particularly focusing on RFID application is continues to progress. A recent study, Yilmaz et al. [45] proposed an ARMOR as an anti-counterfeit mechanism for PUF-based RFID systems. A lightweight three-flights mutual authentication protocol based on the combination of Rabin public-key encryption scheme and PUF technology is proposed to build the anti-counterfeit RFID design.

2.5.2 Cryptographic Key Generation

A unique and device-specific of responses generated from PUFs can be used in cryptographic key generation. Nevertheless, the PUF responses are noisy, hence it requires an error correction code (ECC) to generate an error-free cryptographic key [41]. Figure 2.16 illustrates the procedure of generating an error-free key from a PUF. The procedure consists of two phases; (a) Enrolment, and (b) Reconstruction. The enrolment phase performs once in a controlled and trusted environment. During this phase, the PUF responses, r are measured and extracted. A subset of r, m is fed into the fuzzy extractor which consists of the privacy amplification and ECC blocks. The privacy amplification block (e.g., hash function) is used to increase the entropy of generated key, key. m is encoded using ECC to become a codeword, n. Subsequently, a helper data, h computed as $h = r \oplus n$ is stored in NVM.

In the reconstruction phase, h is retrieved to correct the noisy response, r'. The noisy codeword, n' is computed as $n' = h \oplus r'$ and further decoded using ECC to generate m. Finally, an error-free cryptographic key is generated after the privacy amplification block. As discussed in [30], the BoseChaudhuri-Hocquenghem (BCH) scheme can be used as an ECC for generating error-free cryptographic keys.

The cryptographic key generation procedure as described in Fig. 2.16 can be used to provide trustworthiness to the whole system. Schaller et al. [36] proposed a hardware-assisted solution or hardware-software binding by reusing the available on-chip SRAM in an ARM-based low-end microcontroller as a PUF for firmware protection. As explained in Sect. 2.4.1, SRAM-PUF generates random responses after the power-up process. Therefore, a modification to the boot-loader is required to implement the secret key extraction and decryption functionality when

Fig. 2.16 The procedure of PUF-based cryptographic key generation. (**a**) Enrolment. (**b**) Reconstruction

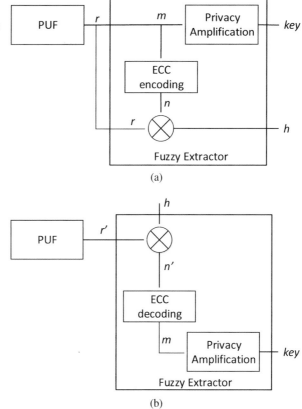

(a)

(b)

the device is powered-up. Figure 2.17 illustrates the proposed hardware-software binding scheme to protect the firmware. The process of generating an error-free cryptographic key is represented by the dashed line box in Fig. 2.17.

Several other potential applications of PUFs such as intellectual property (IP) protection [39], remote activation of ICs [4], hardware-software binding [16], and hardware obfuscation [12, 43] have been proposed. Today, PUF-based hardware security has stepped out of the research labs into next-generation security products. For example, NXP uses PUF as secure key storage to improve the security level of their incoming products [33].

2.6   Conclusion

The rising of interest in trusted computing solutions is driven largely by the emergence of IoT technology and it is becoming more pronounced, nowadays. A generic computing system consists of several layers which are hardware, firmware,

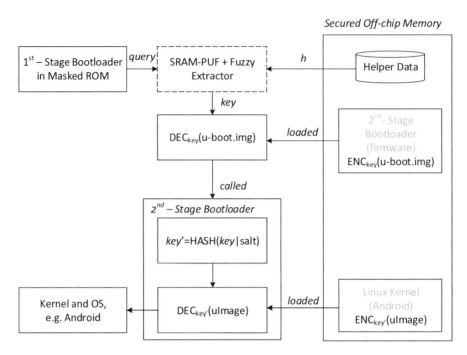

Fig. 2.17 PUF-based hardware-assisted for firmware protection

software (i.e., operating system and application), and data layers. Building security at a software layer could secure the computing system. However, the trustworthiness of the hardware is questionable whether it is genuine or cloned hardware. Hence, an important element of trusted computing is it must be able to provide intrinsically trustworthy to the system. This chapter presented hardware security primitive, known as Physical Unclonable Function (PUF). PUF is a promising technology to provide secure, reliable and trustworthy to any computing systems. PUF extracts a unique hardware characteristic (i.e., device-specific binary response) by exploiting the intrinsic process variations during IC fabrication. A device-specific PUF response can be used as a hardware fingerprint to provide a trusted hardware layer. The development of PUF has evolved following the evolution in semiconductor devices. PUF constructions started off by exploiting intrinsic variations in MOSFET transistors. Recent exploration shows that PUF construction is exploiting intrinsic variations from FinFET transistors and emerging nano-devices such as memristor, CNFET, PCM, etc. PUFs are seen to have several advantages such as energy-efficient, low-overhead, low manufacturing cost, non-programmable, and easy to manage secret keys. Therefore, PUFs have been suggested for many applications such as RFID, IP protection, IC metering, hardware-software binding, hardware obfuscation, etc.

References

1. A. Asenov, A.R. Brown, J.H. Davies, S. Kaya, G. Slavcheva, Simulation of intrinsic parameter fluctuations in decananometer and nanometer-scale MOSFETs. IEEE Trans. Electron Dev. **50**(9), 1837–1852 (2003)
2. M. Bayoumi, A. Dutta, FinFET based SRAM design : A survey on device, circuit, and technology issues, in *IEEE International Conference on Electronics, Circuits and Systems*, pp. 387–390 (2014)
3. Y. Cao, L. Zhang, S.S. Zalivaka, C.h. Chang, S. Chen, CMOS image sensor based physical unclonable function for coherent sensor-level authentication. IEEE Trans. Circuits Syst. **62**(11), 2629–2640 (2015)
4. A. Cui, X. Qian, G. Qu, H. Li, A new active IC metering technique based on locking scan cells, in *IEEE Asian Test Symposium*, pp. 36–41 (2017)
5. J. Delvaux R. Peeters, D. Gu, I. Verbauwhede, A survey on lightweight entity authentication with strong PUFs. ACM Comput. Surv. **48**(2), 26:1–26:42 (2015)
6. Y. Gao, D.C. Ranasinghe, S.F. Al-Sarawi, O. Kavehei, D. Abbott, mrPUF: A novel memristive device based physical unclonable function, in *International Conference Applied Cryptography Network Security*, pp. 595–615 (2015)
7. Y. Gao, D.C. Ranasinghe, S.F. Al-Sarawi, O. Kavehei, D. Abbott, Emerging physical unclonable functions with nanotechnology. IEEE Access **4**(PP), 61–80 (2016)
8. B. Gassend, D. Clarke, M. van Dijk, S. Devadas, Silicon physical random functions, in *ACM Conference on Computer and Communications Security*, pp 148–160 (2002)
9. J. Guajardo, S.S. Kumar, G.J. Schrijen, P. Tuyls, FPGA intrinsic PUFs and their use for IP protection, in *International Conference on Cryptographic Hardware and Embedded Systems*, pp. 63–80 (2007)
10. U. Guin, A. Singh, M. Alam, J. Canedo, A. Skjellum, A secure low-cost edge device authentication scheme for the internet of things, in *International Conference on VLSI Design*, pp. 85–90 (2018)
11. B. Halak, A primer on cryptographic primitives and security attacks, *Physically Unclonable Functions: From Basic Design Principles to Advanced Hardware Security Applications* (Springer International Publishing, 2018), pp. 1–15
12. N.A. Hazari, F. Alsulami, M. Niamat, FPGA IP obfuscation using ring oscillator physical unclonable function, in *IEEE National Aerospace Electronics Conference* (IEEE, 2018), pp. 105–108
13. D.E. Holcomb, W.P. Burleson, K. Fu, Power-Up SRAM state as an identifying fingerprint and source of true random numbers. IEEE Trans. Comput. **58**(9), 1198–1210 (2009)
14. Intel, Intel 22 nm FinFET Low Power (22FFL) Technology: FinFET Technology for the Mainstream (2017). https://newsroom.intel.com/newsroom/wp-content/uploads/sites/11/2017/03/22-nm-finfet-fact-sheet.pdf
15. V.P. Klybik, A.A. Ivaniuk, Use of arbiter physical unclonable function to solve identification problem of digital devices. Autom. Control Comput. Sci. **49**(3), 139–147 (2015)
16. F. Kohnhäuser, A. Schaller, S. Katzenbeisser, PUF-based software protection for low-end embedded devices, in *Trust and Trustworthy Computing*, ed. by M. Conti, M. Schunter, I. Askoxylakis (Springer International Publishing, 2015), pp. 3–21
17. S.T.C. Konigsmark, L.K. Hwang, D. Chen, M.D.F. Wong, CNPUF: A carbon nanotube-based physically unclonable function for secure low-energy hardware design, in *Asia and South Pacific Design Automation Conference*, pp. 73–78 (2014)
18. S.S. Kumar, J. Guajardo, R. Maes, G.J. Schrijen, P. Tuyls, Extended abstract: The butterfly PUF protecting IP on every FPGA, in *IEEE International Workshop on Hardware-Oriented Security and Trust*, pp. 67–70 (2008)
19. J.W. Lee, D. Lim, B. Gassend, G.E. Suh, M. van Dijk, S. Devadas, A technique to build a secret key in integrated circuits for identification and authentication applications, in *Symposium on VLSI Circuits Digest of Technical Papers*, pp. 176–179 (2004)

20. V.V.D. Leest, R. Maes, G.J.S. Pim, P. Tuyls, Hardware intrinsic security to protect value in the mobile market, in *Information Security Solutions Europe*, pp. 188–198 (2014)
21. D. Lim, J.W. Lee, B. Gassend, G.E. Suh, M.V. Dijk, S. Devadas, Extracting secret keys from integrated circuits. IEEE Trans. Very Large Scale Integr. VLSI Syst. **13**(10), 1200–1205 (2005)
22. K. Lofstrom, W.R. Daasch, D. Taylor, IC identification circuit using device mismatch, in *IEEE International Solid-State Circuits Conference*, pp. 372–373 (2000)
23. A. Maiti, P. Schaumont, The impact of aging on a physical unclonable function. IEEE Trans. Very Large Scale Integr. VLSI Syst. **22**(9), 1854–1864 (2014)
24. A. Maiti, V. Gunreddy, P. Schaumont, A systematic method to evaluate and compare the performance of physical unclonable functions, in *Embedded Systems Design with FPGAs*, ed. by P. Athanas, D. Pnevmatikatos, N. Sklavos (Springer, New York, 2013), pp. 245–267
25. M. Majzoobi, F. Koushanfar, M. Potkonjak, Lightweight secure PUFs, in *IEEE/ACM International Conference on Computer-Aided Design*, pp. 670–673 (2008)
26. J. Mathew, R.S. Chakraborty, D.P. Sahoo, Y. Yang, D.K. Pradhan, A novel memristor-based Hardware security primitive. ACM Trans. Embed. Comput. Syst. **14**(3), 1–20 (2015)
27. M.S. Mispan, Towards reliable and secure physical unclonable functions. Ph.d. thesis, University of Southampton (2018)
28. M.S. Mispan, B. Halak, M. Zwolinski, NBTI aging evaluation of PUF-based differential architectures, in *IEEE International Symposium on On-Line Testing and Robust System Design*, pp. 103–108 (2016)
29. M.S. Mispan, B. Halak, M. Zwolinski, Lightweight obfuscation techniques for modeling attacks resistant PUFs, in *IEEE International Verification and Security Workshop*, pp. 19–24 (2017)
30. M.S. Mispan, S. Duan, B. Halak, M. Zwolinski, A reliable PUF in a dual function SRAM. Integration **68**, 12–21 (2019)
31. M.S. Mispan, M. Zwolinski, B. Halak, Ageing mitigation techniques for SRAM memories, in *Ageing of Integrated Circuits* (Springer Nature Switzerland AG, 2020), pp. 91–111
32. B. Narasimham, D. Reed, S. Gupta, E.T. Ogawa, Y. Zhang, J.K. Wang, SRAM PUF quality and reliability comparison for 28 nm planar vs. 16 nm FinFET CMOS processes, in *IEEE International Reliability Physics Symposium (IRPS)*, pp. 2–5 (2017)
33. NXP, Step up security and innovation with next generation SmartMX2 products (2016). https://cache.nxp.com/docs/en/brochure/75017695.pdf
34. R. Pappu, B. Recht, J. Taylor, N. Gershenfeld, Physical one-way functions. Science **297**(5589), 2026–2030 (2002)
35. U. Rührmair, J. Sölter, F. Sehnke, X. Xu, A. Mahmoud, V. Stoyanova, G. Dror, J. Schmidhuber, W. Burleson, S. Devadas, PUF modeling attacks on simulated and silicon data. IEEE Trans. Inf. Forensic Secur. **8**, 1876–1891 (2013)
36. A. Schaller, T. Arul, V. Van Der Leest, S. Katzenbeisser, Lightweight anti-counterfeiting solution for low-end commodity hardware using inherent PUFs, in *Trust and Trustworthy Computing* (Springer International Publishing, 2014), pp. 83–100
37. S.P. Skorobogatov, Semi-invasive attacks-a new approach to hardware security analysis. Tech. rep., University of Cambridge (2005)
38. G.E. Suh, S. Devadas, Physical unclonable functions for device authentication and secret key generation, in *ACM/IEEE Design Automation Conference*, pp. 9–14 (2007)
39. P. Sun, A. Cui, A new pay-per-use scheme for the protection of FPGA IP, in *IEEE International Symposium on Circuits and Systems*, pp. 1–5 (2019)
40. S. Sutar, S. Member, A. Raha, Memory-based combination PUFs for device authentication in embedded systems. IEEE Trans. Multi-Scale Comput. Syst. **4**(4), 793–810 (2018)
41. V. van der Leest, P. Tuyls, Anti-counterfeiting with hardware intrinsic security, in *Design, Automation & Test in Europe Conference & Exhibition*, pp. 1137–1142 (2013)
42. E.I. Vatajelu, G.D. Natale, M.S. Mispan, B. Halak, On the encryption of the challenge in physically unclonable functions, in *IEEE International Symposium on On-Line Testing and Robust System Design*, pp. 115–120 (2019)

43. J.B. Wendt, M. Potkonjak, Hardware obfuscation using PUF-based logic, in *IEEE/ACM International Conference on Computer-Aided Design, Digest of Technical Papers*, pp. 270–277 (2015)
44. Y. Ye, F. Liu, M. Chen, S. Nassif, Y. Cao, Statistical modeling and simulation of threshold variation under random dopant fluctuations and line-edge roughness. IEEE Trans. Very Large Scale Integr. VLSI Syst. **19**(6), 987–996 (2011)
45. Y. Yilmaz, Vh. Do, B. Halak, ARMOR : An anti-counterfeit security mechanism for low cost radio frequency identification systems. IEEE Trans. Emerg. Top. Comput., 1-15 (2020)
46. L. Zhang, Z.H. Kong, C.H. Chang, A. Cabrini, G. Torelli, Exploiting process variations and programming sensitivity of phase change memory for reconfigurable physical unclonable functions. IEEE Trans. Inf. Forensics Secur. **9**(6), 921–932 (2014)
47. S. Zhang, B. Gao, D. Wu, H. Wu, H. Qian, Evaluation and optimization of physical unclonable function (PUF) based on the variability of FinFET SRAM, in *International Conference on Electron Devices and Solid-State Circuits*, pp 1–2 (2017)
48. W. Zhao, Y. Cao, New generation of predictive technology model for sub-45 nm early design exploration. IEEE Trans. Electron Devices **53**(11), 2816–2823 (2006)

Part II
Authentication Protocols

Chapter 3
ASSURE: A Hardware-Based Security Protocol for Internet of Things Devices

Yildiran Yilmaz, Leonardo Aniello, and Basel Halak

Abstract This chapter discusses the design, implementation and evaluation of a hardware-based mutual authentication and the key agreement protocol. The latter combines a lightweight symmetric cipher with physically unclonable functions technology to provide an energy-efficient solution that is particularly useful for Internet of Things (IoT) systems. The security of the proposed protocol is rigorously analysed under various cyberattack scenarios. For overheads' evaluation, a wireless sensor network using typical IoT devices, called Zolertia Zoul RE-mote, is constructed. The functionality of the proposed scheme is verified using a server–client configuration. Then energy consumption and memory utilisation are estimated and compared with the existing solutions, namely the DTLS (datagram transport layer security) handshake protocol in pre-shared secret (PSK) mode and UDP (user datagram protocol). Experimental analysis results indicate that the proposed protocol can save up to 39.5% energy and use 14% less memory compared to the DTLS handshake protocol.

Keywords PUF · Mutual authentication · Key agreement protocol · Scyther · Lightweight symmetric cipher · Energy · IoT security DTLS · UDP

3.1 Introduction

With recent developments in Internet of Things (IoT) systems, it has become possible to design and develop multifunctional sensor nodes that are small in size, come at a low cost, require little power and communicate wirelessly in short distances [1]. The aim of the wireless sensor network is to detect physical and environmental changes such as temperature, sound, pressure and humidity. The

Y. Yilmaz · L. Aniello · B. Halak (✉)
University of Southampton, Southampton, UK
e-mail: yy6e14@southamptonalumni.ac.uk; l.aniello@soton.ac.uk; basel.halak@soton.ac.uk; bh9@ecs.soton.ac.uk

© Springer Nature Switzerland AG 2021
B. Halak (ed.), *Authentication of Embedded Devices*,
https://doi.org/10.1007/978-3-030-60769-2_3

sensor should transmit the sensed data to a central hub in cooperation over the established network. These networks have a wide application area in many sectors, including military, environmental, health, industrial and smart homes. It is worth considering patient-monitoring systems as a concrete example of this technology in use [1]. With this system, patient pulse rate, blood oxygen level, electrical activity in the heart, muscle activation and general kinaesthetic movements may be measured. This makes it possible to monitor the patient's health remotely and intervene in case of an emergency. In this case, robust authentication and secure communication mechanisms are crucial for ensuring that the health readings gathered are trustworthy and for preventing potential attackers from falsifying that information for jeopardising the health of the patient. Two main considerations must be taken into account when providing security solutions for such systems.

1. IoT devices and sensors used for data collection and measurement have limited computing and energy resources; in fact the majority of such devices will be either battery operated (e.g. portable patient-monitoring systems) or rely on energy-scavenging mechanisms (e.g. industrial sensors).
2. IoT devices are typically deployed with no physical protection (e.g. it is feasible or a malicious party to have access to a patient-home monitoring system), which means adversaries can exploit the physical access to such devices to carry well-known invasive and side channel analysis attacks to extract sensitive data or to even clone the whole device.

Security solutions relying on classic cryptographic algorithms [2] can be prohibitively expensive in this case. To address this challenge, a number of solutions have emerged, such as the constrained application protocol (CoAP), which is based on the use of the datagram transport layer security (DTLS) protocol [1, 3, 4]. Other approaches are based on the use of physically unclonable functions (PUF) technology such as [5], which allows for a better physical security. Existing PUF-based protocols, such as those in [6–9], still have a number of limitations, preventing a wider adoption. These include:

1. The need, in some cases, for storing a large number of challenge/response pairs for each PUF at a central verifier for subsequent verification, makes impractical to use this technology for securing networks with large number of nodes.
2. Lack of proper evaluation of the cost associated with the PUF protocols compared to the existing non-PUF solutions (e.g. DTLS).
3. Lack of formal security analysis lack against protocol-level attacks.

This chapter proposes a hardware-based mutual authentication and key agreement protocol called ASSURE (A hardware-baSed SecUrity pRotocol for internEt of things devices). The solution combines a lightweight symmetric cipher with physically unclonable functions technology, in order to achieve more energy efficiency and better physical security. The systematic security analysis of the proposed protocol demonstrates its resilience against a range of attacks, namely man-in-the-middle, eavesdropping, replay, server and client impersonation, de-synchronization, and PUF-modelling attacks. An IoT demonstrator network is constructed for functional verification and costs analysis. The results reveal that the proposed protocol can save up to 39.5% energy and use 14% less memory compared to the DTLS handshake protocol.

3.2 Chapter Overview

The rest of this chapter is organised as follows. Section 3.3 explains the principles of DTLS protocol, PUF and RC5 cipher. Section 3.4 describes the proposed solution. Section 3.5 presents a systematic security analysis performed using the Scyther tool. Section 3.6 describes experimental set-ups and the implementation of the hardware demonstrator. A detailed cost analysis and discussion are presented in Sect. 3.7. Conclusions are drawn in Sect. 3.8.

3.3 Related Background

3.3.1 Physically Unclonable Functions

PUFs can be generally classified as strong or weak constructions. The latter have limited number of response/challenge pairs [10]. In this work, a strong PUF is employed. In this context, this means a PUF with sufficiently large number of challenge/response pairs such that they exceed the number of times a device needs to run the authentication protocol throughout its expected period of usage. The PUF response used in this work is assumed to be stable. This can be achieved using a combination of approaches, such as error correction techniques with the helper data [8, 9], temporal redundancy approaches or the use of emerging error-free designs such as PUF-FSM [11].

3.3.2 Principles Datagram Transport Layer Security

3.3.2.1 Definition

DTLS is a modification of the TLS protocol described in [2]. The main difference between DTLS and TLS is that the former runs on top of UDP (Unreliable Data Transport) connection while the latter runs over TCP (Transmission Control Protocol). Additionally, DTLS has a protection mechanism against denial of service attacks (DoS), which allows out-of-order arrival and retransmission of messages by modifying the handshake header of TLS. There are a number of versions including DTLS 1.0 and DTLS 1.2 published in RFC 4347 and RFC 6347, respectively. The DTLS authentication protocol consists of two phases:

(a) The registration phase relies on the certification authority (CA), an existing infrastructure that is authorised to produce commercial certificates [12]. It include two steps: (1) the server stores the personal certificate S_{cert} issued by CA and the server's private key P_S; (2) the client stores the personal certificate C_{cert} issued by CA and the client's private key P_C.

(b) The handshake phase has three objectives. First, it allows the communicating parties to agree on a particular cipher suit to use, which facilitates interoperability between different implementations of the protocol. A cipher suit typically specifies one algorithm for each of the following tasks: key exchange, symmetric key encryption, message authentication and a mode of operation. Second, it allows the communicating parties to perform mutual authentication using digital certificates, to learn each other's keys and verify each other's identities. Third, it allows the establishment of a shared secret key to be used in subsequent secure communication [2]. In summary, the protocol version and the selection of encryption algorithm, that is cipher suite, are negotiated in this handshake process. By receiving and validating the digital certificates, both parties, server and client, perform authentication mutually, as described in [13].

3.3.2.2 DTLS Protocols

Traditional designs of DTLS use RSA-based authentication as the default-operating mode. However, the significant computational overheads of this asymmetric cryptosystem prevent its use in resource-limited devices [3], unless they have smaller standardised trusted certificates and energy-efficient protocols with fewer interactions [14]. Therefore, for such systems, alternative DTSL modes can be used, more specifically pre-shared secret key (PSK) and the raw-public key (RPK) implementations. The PSK-mode DTLS requires the client and the server to generate and share a cryptographic key before the start of a session. On the other hand, the RPK mode uses public-key-based authentication [15]. These raw-public keys must be kept and protected by a trusted entity. This mode assumes that the communicating devices know each other's public keys [16]. Both PSK and RPK modes require six interactions between the client and the server for mutual authentication and for exchanging and agreeing of security parameters.

Other works in [17, 18] explored the use of ECC cryptography, which was proposed for the CoAP [19]; their results, however, have not indicated a great deal of improvement compared to the original DTLS. In this chapter, the PSK (pre-shared key) mode was chosen for comparative cost analysis as it is deemed one of the most efficient implementations of DTLS [20].

3.3.3   RC5 Algorithm for Resource-Limited Environments

RC5 depicted in Fig. 3.1 is a symmetric encryption algorithm, which can operate with word lengths of 16 to 64 and 128 bits. The key size and number of rounds that are fixed in traditional encryption algorithms such as AES and IDEA can be variable in the RC5 algorithm. The inherent configurability of RC5 makes it particularly attractive to lightweight applications, as it provides the flexibility of adjusting the complexity of the cipher, hence the computation costs, according to the required level of security [21]. The RC5 algorithm has three basic components, namely a key expansion algorithm, encryption algorithm and decryption algorithm, which will be discussed below.

3.3.3.1   RC5 Notation

The RC5 algorithm is commonly represented along with parameters in the following way: RC5 – w/r/b. For example, a combination of RC5–32/16/18 shows 32-bit word length, 16-round and 18-byte (80-bit) key. For optimal IoT security, the recommended parameter selection [21] is a 128-bit key, 64-bit block size and 12 rounds, as adopted in this work (Table 3.1).

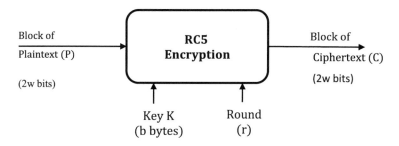

Fig. 3.1 RC5 Encryption block diagram

Table 3.1 Parameters of RC5 encryption/decryption

Parameters	Description
w	Word length. Each word contains $u = (w/8)$ 8-bit bytes. RC5 encrypts binary word blocks; plain text and encrypted text are 2w bits long. Word length could be 16, 32 or 64 bits
r	Number of rounds
b	Key length in byte. Key length could be 0 to 255 bits

3.3.3.2 Encryption Process

The encryption process, shown in Fig. 3.2, includes simple operations such as xor, addition and rotation. It is-considered that the input block is in the form of two w-bits. It is also assumed that the key block is generated and the $S[0 \ldots t-1]$ array is completed. The example pseudocode for the encryption process is given below:

Pseudocode 1	Encryption process
Step 1	$A = A + S[0]$;
Step 2	$B = B + S[1]$;
Step 3	for $i = 1$ to r do
Step 4	$A = ((A \oplus B) <<< B) + S[2*i]$;
Step 5	$B = ((B \oplus A) <<< A) + S[2*i+1]$;

In the end, the output of the process is the-encrypted A and B.

3.3.3.3 Decryption Process

A pseudocode for the decryption process is given below:

Pseudocode 2	Decryption process
Step 1	for $i = r$ downto 1 do
Step 2	$B = ((B - S[2*i+1] >>> A) \oplus A$;
Step 3	$A = ((A - S[2*i] >>> B) \oplus B$;
Step 4	$B = B - S[1]$;
Step 5	$A = A - S[0]$;

3.3.3.4 Key Expansion in RC5 Algorithm

This process expands the key using the following four operations:

A) Magic constants generation

The key expansion algorithm uses golden ratio $(\emptyset)$ and base of the natural logarithm (e) to generate two constants P_w, Q_w. These constants are generated as follows:

$$P_w = \text{Odd}\left((e - 2) \, 2^w\right)$$

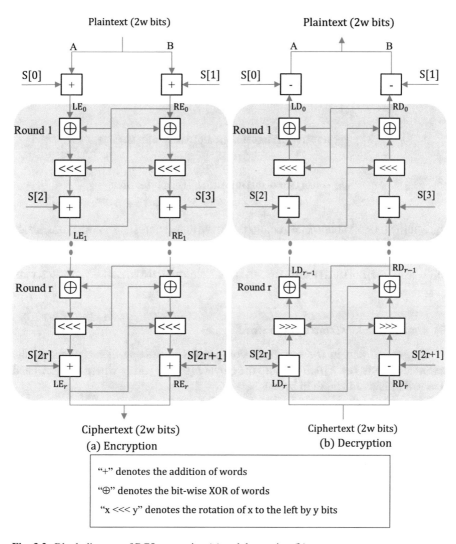

Fig. 3.2 Block diagram of RC5 encryption (**a**) and decryption (**b**)

$$Q_w = \text{Odd}\left((\varnothing - 2)\, 2^w\right)$$

where

Odd(x) refers to the odd-integer closest-to x.
e = 2.718281828 . . . (base of the natural logarithm)
$\varnothing$ = 1.618033988 . . . (golden−ratio)

In the case where w is 16, 32 and 64, the following constants are produced:

$$P_{16} = 1011011111100001 = b7e1$$

$$Q_{16} = 1001111000110111 = 9e37$$

$$P_{32} = 10110111111000010101000101100011 = b7e15163$$

$$Q_{32} = 10011110001101110111100110111001 = 9e3779b9$$

$$P_{64} = 1011011111100001010100010110001010001010111011010010101001101011 = b7e151628aed2a6b$$

$$Q_{64} = 1001111000110111011110011011100010111111101001010011111100000010101 = 9e3779b97f4a7c15$$

B) Converting key from byte to word

The second part in the key expansion process is to copy the sequence of the generated b byte key $K[0..b-1]$ into the c byte $L[0..c-1]$ array where $c = [b/u]$ and $u = w/8$ (u is word length in bytes).

The example pseudocode for key conversion process is given below:

Pseudocode 3	Key conversion process
Step 1	$c = [max\ (b,1)/u]$
Step 2	for $i = b - 1$ downto 0
Step 3	$\quad L[i/u] = (L[i/u] <<< 8) + K[i]$

C) The initialisation of the key array S

The third part in the key expansion process is performed as follows. The first element of the S array is filled using P_w, then the rest of the elements of the S array from 1 to $t-1$ are filled by the addition of Q_w to the previous element of the S array in the loop. The example pseudocode is given below:

Pseudocode 4	Key initialisation process
Step 1	$S[0] = P_w;$
Step 2	for $i = 1$ to $t - 1$
Step 3	$\quad S[i] = S[i-1] + Q_w$

D) **Mixing within the key**

The final part in the key expansion process is the mixing of the key using the S and L arrays created previously. Because the lengths of the S and L arrays are different, the process is applied three times. The example pseudocode is given below:

Pseudocode 5	Key mixing process
Step 1	$i = j = 0$
Step 2	do 3*max (t, c) times:
Step 3	$S[i] = (S[i] + \text{reg}1 + \text{reg}2) <<< 3$
Step 4	$\text{reg}1 = S[i]$
Step 5	$L[j] = (L[j] + \text{reg}1 + \text{reg}2) <<< (\text{reg}1 + \text{reg}2)$
Step 6	$\text{reg}2 = L[j]$
Step 7	$i = (i + 1) \bmod t$
Step 8	$j = (j + 1) \bmod c$

3.4   ASSURE – Protocol Description

3.4.1   Specifications

The proposed solution combines PUF technology with the RC5 cipher to produce a new protocol that has the same security properties of DTSL against protocol attacks, while at the same time consuming less energy. In addition to this, it provides security against physical attacks, thanks to PUF cryptographic primitive [22]. The protocol satisfies the following requirements:

- *Mutual authentication*: Both parties' (a client and a server) communication over a network must authenticate each other.
- *Peer-to-peer key agreement*: Two IoT devices authenticated by the same server must be able to agree on the key for further secure communication.
- *Client Unclonability*: When the attacker obtains all the design features of the IoT device, the security of the system should not be compromised by means of cloning the device.
- *PUF model on the verifier:* A software model of the PUF model is used at the server side to reduce the cost of storing a large number of challenge/response pairs.

It is worth emphasising that the choice of the light cipher was driven by cost implications; however, the presented protocol can be used with any symmetric cipher.

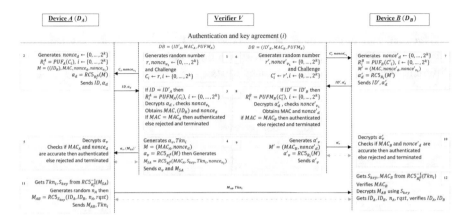

Fig. 3.3 ASSURE protocol description

3.4.2 Operation Principles

The protocol described here assumes a network of three nodes: a resource-rich server that acts as a verifier (V) and two lightweight devices A and B (i.e. provers), represented by D_A and D_B. Each device includes a PUF, and the verifier includes a software model of the PUF of each node in the system. These models are generated using machine learning algorithms. The initial communication is established using a UDP connection. The protocol consists of three phases, namely the registration phase, the verification phase and the peer-to-peer key agreement phase, as shown in Fig. 3.3. Table 3.2 provides a description of the symbols used within the protocols

a) Registration Phase

This step is performed for each new device joining the network. In the scenario under consideration, the verifier (V) obtains PUF models, that is $PUFM_A$ and $PUFM_B$,using a neural network algorithm and stores them with the MAC address and IDs of devices (A and $B)$ into the database, for example

$$DB = (ID\prime_A, MAC_A, PUFM_A).$$

b) Authentication Phase

The first three flights in Fig. 3.3 are for device–server mutual authentication, and the fourth and last flight is for device–device key agreement. The authentication processes of devices A and B are carried out with similar actions, with the exception of the third flight. The authentication steps of device A are detailed below:

Table 3.2 List of symbols

Parameter	Description
ID'	The device's identifier
V	The verifier to which the device wants to connect (server)
k	The bit length of the PUF
r	A random number generator defined in the set Z_n^+
$nonce$	Random number
C, R	Challenge and response
$RC5$	Rivest cipher
$RC5_{key}$	Subscript of RC5 indicating the key used (128 bit)
DB	Database
T_p	The time period in which the token(Tkn) may live.
S_{key}	Symmetric key to be agreed and utilised for creating a session between D_A and D_B.
Tkn	Token for sharing S_{key} between D_A and D_B
M	Message containing confidential data
Mac	Media access control address, which is a unique identifier assigned to a network interface controller (NIC) of each device within the network

Step 1. The verifier generates a random number and $nonce_{v_1}$ $0 \leq r \leq 2^k$, $r \in R$, normalises the challenge $C_i = r, 0 < i \leq 2^n$, $i \in Z^+$ and sends the challenge (C_i) and $nonce_{v_1}$ to device A.

Step 2. Device A receives the challenge (C_i) and generates the response $R_i^A = PUF_A(C_i)$. Afterwards, device A computes $\alpha_d = RC5_{R_i^A}(M)$, where $M = \left((ID_B), MAC, nonce_d, nonce_{v_1}\right)$, and sends it along with its ID to the verifier.

Step 3. Device A is checked by looking at its ID in the database of the verifier. The verifier computes the response $R_i^A = PUFM_A(C_i)$. Then, the verifier decrypts α_d and obtains $(ID_B), MAC, nonce_{v_1}$ and $nonce_d$. If ID_B exists in message (α_d), that means that device A wants to communicate with device B. The verifier checks the authenticity of the MAC from the database and verifies the $nonce_{v_1}$. The freshness control is based on $nonce_{v_1}$. This implies that the received information is recent. If those checks are successful and the data is fresh, the device is authenticated; otherwise, the device is rejected.

Step 4. After the authentication of device A has been successfully completed, the verifier computes and sends $\alpha_v = RC5_{R_i^A}(MAC_A, nonce_d)$ to be authenticated by A. The token (Tkn_i) is generated by the verifier such that $Tkn_i = RC5_{R_i^B}(ID_A, MAC_B, T_p, S_{key})$. V then generates a random $nonce_{v_2}$ and sends it with Tkn_i in an encrypted form to A such that $M_{SA} = RC5_{R_i^A}\left(MAC_A, S_{key}, Tkn_i, nonce_{v_2}\right)$. Transmitting M_{SA^*} containing Tkn_i is optional and occurs when device A wants to communicate with device B.

Step 5. The device decrypts α_v and checks MAC_A. If the check is successful and $nonce_d$ is fresh, the server (verifier) is authenticated.

The mutual authentication steps between B and V are processed in a similar way as follows:

Step 6. The verifier generates and sends a random number $nonce'_{v_1}$ and normalised challenge C'_i to device B.

Step 7. Device B generates the response $R_i^B = PUF_B(C'_i)$ then computes α'_d and sends it along with its ID' to the verifier.

Step 8. Device B is checked by looking at its ID' in the database of V. Afterwards, V computes the response $R_i^B = PUFM_B(C'_i)$, then decrypts α'_d and obtains $MAC, nonce'_{v_1}$ and $nonce'_d$. V checks the authenticity of the MAC from the database and verifies the $nonce'_{v_1}$.

Step 9. If those checks are successful, V computes and sends $\alpha'_v = RC5_{R_i^B}(MAC_B, nonce'_d)$ to B.

Step 10. Device B decrypts α'_v and checks MAC_B and $nonce'_d$, then authenticates the server.

c) **Peer-to-Peer Key Agreement Phase**

This is an optional phase of the protocol, which only takes place if two nodes in the network want to establish a secure communication, hence the need to agree on a shared secret key for encryption.

Step 11. Device A generates a request ($rqst$) and a random nonce (n_A) and sends them with ID_A, ID_B in an encrypted form in a message such that $M_{AB} = RC5_{S_{key}}(ID_A, ID_B, n_A, rqst)$, along with Tkn_i to device B.

Step 12. Device B obtains R_i^B, which is already used in the authentication phase, using its PUF_B. B decrypts the Tkn_i using R_i^B and obtains the secret key S_{key}, MAC_B and then verifies MAC_B. It then decrypts the M_{AB} using secret symmetric key S_{key} and obtains the ID_A, ID_B, n_A and $rqst$. If obtaining S_{key} and decrypting M_{AB} is successful, then device B verifies whether ID_B, MAC_B are correct or not. If those verifications fail, B terminates the process. Otherwise, the key agreement is considered complete and device B can communicate and share resources securely with device A using S_{key}. In the case of reconnection, device B will receive the same challenge from the server and generate the same response during the current authentication and key agreement session. Consequently, B knows which response to use to decrypt the token received from A.

3.5 Security Analysis

This section describes the security analysis of the protocol for validating the proposed solution. First, a system model and an attacker model are characterised to define the security properties provided by the proposed protocol – namely *mutual authentication, resistance to man-in-the-middle and eavesdropping attacks, resistance to replay attacks and resistance to device impersonation and cloning,*

Network Model

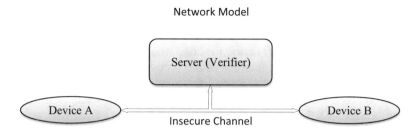

Fig. 3.4 System model for resource-constrained IoT devices

server impersonation and de-synchronisation attacks. These properties are validated using an automatic tool for the analysis of security protocols named Scyther. This section also analyses the security of the proposed protocol against PUF modelling attacks.

3.5.1   System Model

In the system model depicted in Fig. 3.4, a number of parties execute a generic authentication and key agreement protocol by exchanging messages with the aim of proving each other's identity. The protocol specification defines what *roles* are involved in each execution of the protocol itself. For example, an authentication protocol usually includes a *prover* role for the party to be authenticated and a *verifier* role for the party verifying the prover's identity. The specification also defines what messages are sent and received and in which order, their expected content and the state of each involved role, that is what the variables are in the local memory of each role and how they change upon receiving new messages.

There can be more instances of the same protocol being executed at the same time, and the same party can participate in more instances simultaneously. The proposed authentication protocols allow each party to assume the verifier role with two distinct provers shown in Fig. 3.4. Subsequently, the verified nodes can run a peer-to-peer key agreement protocol, agree on a key and start exchanging messages securely.

3.5.2   Attacker Model

In the attack scenario, an attacker is a party potentially playing several roles in different instances of the same authentication and key agreement protocol. This party attempts to deceive honest parties by behaving in compliance with the protocol with the aim of breaking security properties. Relevant examples of attack strategies are described below;

In *a man-in-the-middle (MitM) attack*, the attacker operates in between two honest parties that are running an instance of the protocol. In this type of attack, the attacker can intercept, alter and relay the messages exchanged by the honest parties without them being aware of the intrusion. By operating in the middle, secretly and with the possibility of modifying to some extent the content of intercepted messages, the attacker can break the protocol.

In *an eavesdropping attack*, the data transmitted over a network or channel is listened without permission by malicious third parties to gain access to private information. It is possible that the sensitive data, for example the key, can be obtained and exploited for malicious purposes.

In *a replay attack*, the attacker first picks valid messages exchanged between honest parties in some instance of the protocol, then reuses these messages in a different instance to bypass identity controls and pose as a legitimate party.

In *a device impersonation attack*, the adversary attempts to deceive an authenticator server into accepting a fake device as valid.

In *a cloning attack*, the attacker tampers with a device to retrieve its secret data and identity; these are subsequently used to create a fraudulent device, which can be falsely admitted.

In *a server impersonation attack*, a powerful adversary who aims to obtain the internal state of a device tries to emulate the current server and portray itself as a valid server.

In *a de-synchronization attack*, which is a form of denial of service attacks, the adversary intentionally disrupts the synchronisation of the server with the device, which can impede the authentication process.

In *a PUF modeming attack*, the attacker aims to create a software model of the PUF that mimics its behaviour to high level of accuracy. This goal is achieved by collecting a large number of challenge/response pairs (CRPs) and feeding this into a machine learning algorithm. Once a model of the PUF is obtained, the attacker can perform other attacks such as device cloning or server spoofing. In this work, the attacker is assumed to have access to the communication channel and to the primary input and output of the IoT device.

3.5.3   Security Properties

The ASSURE protocol archives the following security properties:

- *Mutual authentication*: In every instance of the authentication protocol, if both participating parties are honest, then each party correctly verifies the identity of the other. Otherwise, if any participating party behaves maliciously, then the authentication process fails.

- *Resistance to man-in-the-middle attack*: The attacker cannot break the mutual authentication property and peer-to-peer key agreement by running an MitM attack.
- *Resistance to eavesdropping attack*: The attacker cannot compromise the security of the transmitted sensitive data during the protocol execution by eavesdropping on the communication channel between client and server.
- *Resistance to replay attack*: The attacker cannot break the mutual authentication property and key agreement by running a replay attack.
- *Resistance to de-synchronisation*: The Nisynch (non-injective synchronisation) part of the protocol definition in Scyther verified that the authentication and key agreement protocol is not vulnerable to a de-synchronization attack.
- *Resistance to server impersonation*: The adversary cannot reveal the internal state of the device due to PUF and, thus, cannot emulate the proper server.
- *Resistance to device impersonation and cloning:* The proposed protocol has resistance to device impersonation and cloning attack due to the use of PUF technology.
- *Resistance to model-building attacks:* The proposed protocol prevents adversaries from collecting large number of challenge/response pairs (CRP); this is achieved because the output of the PUF is only used internally in the device and never transmitted to other nodes. Therefore, an adversary who can eavesdrop on the communication link or even have physical access to the primary input/output of the device cannot collect sufficient CRP to construct a software model using machine learning algorithms.

3.5.4   Scyther Tool

Scyther is a publicly available tool responsible for formally analysing security protocols. It is assumed that all cryptographic algorithms are fault free, that is an attacker cannot derive any information from encrypted messages without having the encryption key. In order for protocols to be verified, they must be defined using a custom C-like programming language, as advised in Sect. 3.5.6. Each protocol defines the involved roles, the local variables of each role (i.e. its private variables), the content and order of exchanged messages and what specific security properties must be formally verified. Those security properties are specified through so-called *claim events*. A *secrecy claim* of a variable holds if each time an honest party communicates with honest parties, the variable value will not be known to the adversary. A *non-injective synchronisation claim* – also known as Nisynch in Scyther jargon – states that everything intended to happen in the protocol definition also happens in its execution, that is all received messages were sent by the communication partner and all sent messages have been received by the communication partner. Furthermore, *commit claims* are used to require that parties agree on the values of a set of variables.

3.5.5 Motivation for Using Scyther

Security validation for protocols is a challenging task, because Scyther is a security validation tool, which has a specific language named SPDL (stands for Security Protocol Description Language) for modelling communication structure and flow of a security protocol. With this language, the protocols, roles and send/receive events can be defined and expressions for user-defined functions, encryption and hash functions can be represented. This language provides much more understandable and accurate definitions than the traditional mathematical validation method, such as BAN logic [23, 24]. Vulnerabilities and security claims of the protocols defined in this language are automatically approved by trying all protocol attacks against. As stated in [25–27], the performance of Scyther has been broadly examined and compared to the performance of previous protocol verification tools. A significant improvement in performance was found compared with Avispa tools [28]. The Proverif tool [29] has similar performance, but its -syntax is more complicated compared to that of Scyther Therefore, Scyther is chosen in this work for its good performance and ease of use.

3.5.6 Defining the Proposed Protocol in Scyther

Listing 3.1 shows the definition of the proposed protocol using the programming language of Scyther.

Three roles are defined: the devices, that is `role A and B`, and the verifier, that is `role V`. The local status of each part is defined as a set of variables: the nonces `nonced, noncev, noncev2 and nA`, the challenge `Ci` and the response `Ri`.

Listing 3.1 Model of the Proposed Protocol Process in Scyther Jargon

```
//proposed protocol description
const PUF: Function;
protocol proposedProtocol(A,B,V)
{
        role A
        {
                fresh nonced: Nonce;var noncev: Nonce;var Ci: Nonce;
                var Ri: Nonce;var tokeni;var sKey;var noncev2: Nonce;
                fresh nA: Nonce;fresh MACa;fresh rqst;

                recv_1(V, A, Ci, noncev);
                match(Ri, PUF(k(A, V), Ci));
                send_2(A, V, A, {MACa,B,nonced,noncev}k(Ri));
                recv_3(V, A, {MACa,nonced}k(Ri),{MACa,sKey,tokeni,
                noncev2}k(Ri));
```

(continued)

```
        //key agreement
        send_7(A,B, tokeni, {A,B, nonceA,rqst}k(sKey));
        claim_A1(A, SKR, Ri);
        claim_A2(A, SKR, sKey);
        claim_A3(A, Nisynch);
    }
    role B
    {
        fresh nonced: Nonce;var noncev: Nonce;var Ci: Nonce;
        var Ri: Nonce;var sKey;var tokeni;var nA;
        fresh nonceB: Nonce;fresh MACb;var rqst;

        recv_4(V, B, Ci, noncev);
        match(Ri, PUF(k(B,V), Ci));
        send_5(B, V, B, {MACb, nonced,noncev}k(Ri));
        recv_6(V, B, {MACb, nonced}k(Ri));
        //key agreement
        recv_7(A,B, tokeni, {A,B, nA,rqst}k(sKey));
        claim_B1(B, SKR, Ri);
        claim_B2(B, Nisynch);
    }
    role V
    {
        fresh noncev: Nonce;fresh Ci: Nonce;var nonced: Nonce;
        var Ri: Nonce;fresh tokeni;fresh timep: Nonce;fresh sKey: Nonce;
        fresh noncev2: Nonce;var MACa;fresh MACb;
        //authentication of A
        send_1(V, A, Ci, noncev);
        match(Ri, PUF(k(A,V), Ci));
        recv_2(A, V, A, {MACa,B, nonced,noncev}k(Ri));
        match(tokeni, {A,MACb, timep, sKey}k(Ri));
        send_3(V, A, {MACa, nonced}k(Ri),{MACa,sKey,tokeni,
        noncev2}k(Ri));
        //authentication of B
        send_4(V, B, Ci, noncev);
        match(Ri, PUF(k(B,V), Ci));
        recv_5(B, V, B, {MACb, nonced,noncev}k(Ri));
        send_6(V, B, {MACb, nonced}k(Ri));


        claim_V1(V, SKR, Ri);
        claim_V2(V, SKR,tokeni);
        claim_V3(V, SKR,sKey);
        claim_V4(V, Nisynch);
    }
}
```

The device and the verifier generate one nonce each, `nonced` and `noncev`, respectively. The challenge `Ci` is generated by the verifier as it were a nonce. The response `Ri` derives from the `PUF` function declared at the beginning, known to both the device and the verifier. Each of these variables is either declared as `fresh`, meaning that it is randomly generated by the party, or as `var`, meaning that it gets a value, which is assigned to it upon the reception of a message. Messages are sent using the `send` event and received through the `recv` event. The first parameter of the `send` sets the identity of the sender; the second specifies the identity of the destination. The same may be applied to the `recv` event. The other parameters define the content of the message, that is what variables are included in the message. For the `send`, all variables must already have an assigned value. For the `recv`, the local variables included in the content are assigned the values provided in the corresponding `send` event. The `match` event assigns to the variable put as the first parameter the value specified as the second parameter. In this case, the output of the `PUF` functions are computed on the secret key shared between the device and the verifier, that is `k (A, V)`, and on the challenge `Ci`. The `claim` events at the end of each role define the first six security properties for verification, as was already explained in Sect. 3.5.3.

All the above definitions aim to validate the security of the proposed protocol against protocol attacks. The following section will validate the security of the proposed protocol against the model-building attack in particular. Subsequently, proof of security against all the other attacks listed above is discussed in Sect. 3.5.8.

3.5.7   Proof of Security Properties

As shown in Fig. 3.5, upon running the Scyther tool on the proposed protocol, which was itself reported in Sect. 3.5.6, the output shows that all the itemised claims were verified. This means that if the device and the verifier are honest, they mutually agree on what messages are exchanged, as well as their order and content (Nisynch and Commit claims), regardless of the potential actions of the attacker. This guarantees *mutual authentication* and *resistance to the MitM attacks, as well as the eavesdropping and de-synchronization attacks.* If the same nonce is never used more than once, *the resistance to the replay attacks* is ensured as well. Considering that the response R_i is kept secret between the device and the verifier guarantees that the content of the last two exchanged messages cannot be decrypted. Nisynch claim of the protocol was verified by Scyther, that is the device never loses synchronisation with the server because its secrets and public data are freshly generated for each protocol cycle. The de-synchronization attack cannot be performed because listening or stopping any link in the protocol cannot break the synchronisation of data between the communication parties.

The CRP behaviour of the PUF collapses severely if an invasive attack is executed against the device. This PUF property is evidence of tamper resistance [10]. It is impractical to predict the R_i to C_i without accessing the corresponding PUF. The device responds with encrypted text to the server query using the newly generated nonce ($nonce_d$) and response (R_i). Thus, it is impossible for the attacker to generate the correct encrypted text using the challenge and nonce. Consequently, PUF protects sensitive secrets in the device and the encryption prevents the collection of PUF responses from the communication channel. Hence, it is impossible to *impersonate and clone the device.*

Regarding *server impersonation*, it is not possible for a fake server to correctly generate the encrypted message α_v because it depends on the PUF response, the

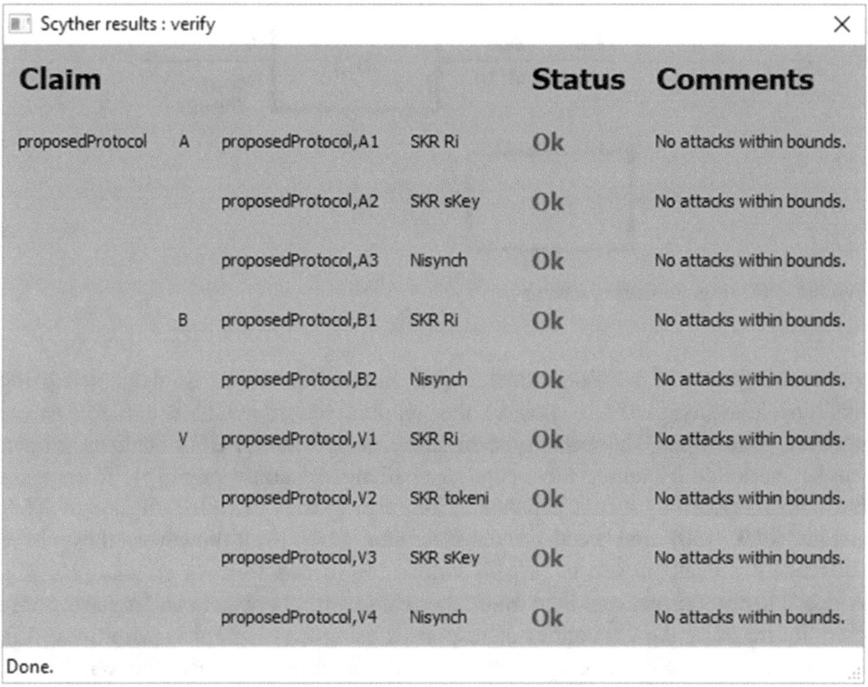

Scyther results : verify					✕
Claim				**Status**	**Comments**
proposedProtocol	A	proposedProtocol,A1	SKR Ri	Ok	No attacks within bounds.
		proposedProtocol,A2	SKR sKey	Ok	No attacks within bounds.
		proposedProtocol,A3	Nisynch	Ok	No attacks within bounds.
	B	proposedProtocol,B1	SKR Ri	Ok	No attacks within bounds.
		proposedProtocol,B2	Nisynch	Ok	No attacks within bounds.
	V	proposedProtocol,V1	SKR Ri	Ok	No attacks within bounds.
		proposedProtocol,V2	SKR tokeni	Ok	No attacks within bounds.
		proposedProtocol,V3	SKR sKey	Ok	No attacks within bounds.
		proposedProtocol,V4	Nisynch	Ok	No attacks within bounds.
Done.					

Fig. 3.5 Scyther verification results

MAC and the random nonce $nonce_d$ value. Furthermore, the fake server cannot identify a device from its *MAC* with relation to its corresponding PUF. The *MAC* address is never exposed since it is always sent in an encrypted way on the channel.

Moreover, the proposed protocol is resilient to *model-building attacks*, which is achieved by hiding the responses sent in the communication channel. The response is masked by using the RC5 algorithm, that is $\alpha_v = RC5_{R_i}(MAC_A, nonce_d)$. In this way, the real response is protected to prevent an adversary from collecting all CRPs and building a model of the PUF, as proven in Sect. 3.5.8.

Thus far, all of the above statements have validated the security of the proposed protocol against protocol attacks and physical invasive attacks, including model-building attacks. With respect to the capabilities of the proposed protocol and its security features, more details will be given in the next section when it is compared to other PUF-based protocols.

3.5.8 Model-Building Resistance

Exposing large number of challenges and corresponding responses of the PUF in the communication channel makes classical PUF-based authentication protocols

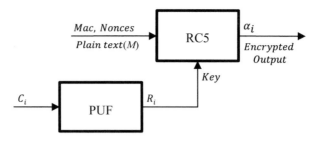

Fig. 3.6 PUF response-hiding scheme

vulnerable to model-building attacks. PUF modelling can be implemented using machine learning methods such as the support vector machine and the neural network algorithm. The behaviour of most delay-based PUFs, such as arbiter, can be modelled by taking advantage of challenge/response pairs [5]. To increase the machine learning attack resistance, initial proposals included the use of XOR Arbiter PUF [30] and Feed-Forward Arbiter PUF [31]; however, these have subsequently been proven to be still vulnerable to this form of attacks [10, 32]. A possible countermeasure is to break the relationship between challenge/response pairs by masking the challenges or responses using cryptographic primitives [33]. Che et al. [34] used a cryptographic hash function to mask the challenges. Response obfuscation has also been proposed [32]. Another approach proposed by Barbareschi et al. [33] employs a PUF model-masking scheme using AES encryption. In their scheme, the R_i is generated by stimulating the Anderson PUF with the C_i; then the C_i is encrypted with AES encryption using the response (R_i) as a key. The encrypted output $E_{R_i}(C_i) = R_i'$ and the C_i pairs (C_i, R_i') are then used in an authentication protocol. The security encryption algorithm prevents adversaries from obtaining PUF responses from the encrypted outputs [6]. Although the use of AES encryption can completely remove the risk of modelling attacks given its confusion and diffusion properties [6], this is a costly approach for resource-constrained devices [35]. Therefore, in this work, the lightweight cipher (RC5) is used to prevent PUF modelling attacks.

The approach used in this work is shown in Fig. 3.6. The response (R_i) generated by the PUF is used as the encryption key. Then, the output (α_i) of the cipher is the RC5 encrypting of the plaintext M, which includes the *Mac* and the *nonces*. The *nonces* are generated by a 40-bit pseudo random number generator (PRNG). The remainder of this section explains how the proposed approach prevents modelling attacks, assuming the adversary has access to the challenge and the output of the RC5 encryption.

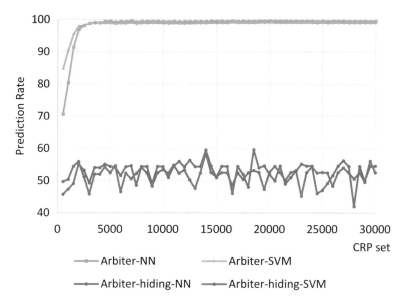

Fig. 3.7 ML-attacks (SVM, NN) on Arbiter PUF and its response-hiding scheme

3.5.8.1   Test Vector Generation and Machine Learning

In the basic PUF-based authentication protocol, the challenge/response pairs are sent in clear between the client and server. Therefore, if an attacker can collect a sufficient number of CRPs, they will be able to create a model with machine learning algorithms. In this work, two experiments are performed. The first aims to demonstrate the possibility of cloning a PUF using machine learning algorithms. The second aims to demonstrate that the ASSURE protocol is secure against this type of threat. In this work, we have used two designs, an arbiter-PUF [36] and a TCO-PUF [37]. Both take a 32-bit challenge as an input and generate a 1-bit response. Both designs have been implemented using CMOS 65 nm technology. A 32000 challenge/response pairs (C_i, R_i) are collected in each case from. Artificial neural network and support vector machine algorithms have been used to carry out the modelling attack.

3.5.8.2   Model-Building Attack Results

The results from Figs. 3.7 and 3.8 demonstrate that both PUF designs considered here can be modelled to a high level of accuracy, 99.5% and 98.4%, if the adversary has access to their challenge/response pairs.

On the other hand, if the response-masking scheme proposed above is used, modelling attacks become unfeasible, as is explained below. The adversary is

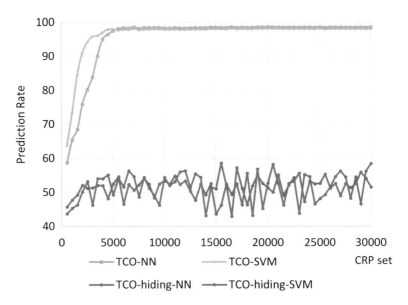

Fig. 3.8 ML-attacks (SVM, NN) on TCO PUF and its response-hiding scheme

assumed to have access to the PUF (C), that is $C = b_1 \cdots b_k$, and the encrypted output (α_i), that is $\alpha = o_1 \cdots o_m$, wherein $\alpha_d = RC5_{R_i^A}(M)$. M consists of 80-bit *nonces* and 48-bit *Mac*. R is the PUF responses used to generate the key.

Figure 3.7 and Fig. 3.8 also depict the prediction results of the response-hiding scheme using both types of ML-attack (SVM and NN). These results are based on the challenge (C_i) and the first 5 bits (o_1, o_2, o_3, o_4, o_5) of the encrypted output o_i. For each challenge (C_i) and o_i bit pairs, the prediction accuracy on average for the Arbiter response-hiding scheme is 52.6% when using ANN-based attack and 51.9% when using SVM-based attack. Similar results are obtained in the case of TCO PUF.

3.6 Experimental Analysis Method

This section outlines the experimental set-ups and explains the metrics used to evaluate the energy and memory-related costs of the proposed protocol.

3.6.1 The Purpose of the Experiment

The purpose of these experiments is twofold. The first is to verify the functionality of the proposed solution; this is achieved by constructing a wireless network that has a server and two clients, as depicted in Fig. 3.4. The second goal of the experiments is

Table 3.3 Power breakdown of the Zolertia Zoul RE-mote node

		Current	Min(sleep)
Computation	CPU	0.6 mA	1.3microA
Radio	LPM	150 nA	–
	Tx(transfer)	24 mA	–
	Rx(listen)	20 mA	–

to evaluate the energy consumption and memory utilisation of the proposed scheme compared to existing solutions, namely the PSK-DTLS and UDP protocols. It is worth noting that the following assumptions are made in this work:

- The means of communication is a wireless link.
- The sensor devices can be deployed in a publicly accessible environment.
- All the sensor devices deployed in the network are resource constrained in terms of energy, memory and computation.
- The central device, which is the server, is comparatively a resource-rich device in terms of energy, memory and computation.

3.6.2 Experimental Set-up

To conduct this experiment, the Zolertia Zoul RE-mote devices, a computer and the Contiki operating system were utilised. The Contiki operating system running on VMware with a 10 GB hard disk and 2 GB of RAM was used for the experimental set-up. In the case of hardware implementation, three Zolertia Zoul devices, equipped with 512 KB flash memory (ROM) and 32 KB RAM, were used, with one serving as the server and the other two as the clients. The devices were programmed with Contiki and energy measurements were estimated by using energest module in the same tool. Table 3.3 presents the current consumption details of the CPU and radio in active and sleep modes derived from the Zoul node datasheet [38]. The operating voltage was 3.4 V. The details that are used to calculate the energy measurement of the specified protocols are in Sect. 3.7.4.

The Zolertia Zoul device was chosen for this work as it is one of the standard IoT devices used in practical applications such as smart homes [38].

The Contiki OS is a well-known open source operating system developed by Adam Dunkels under the C programming language [39]. It was designed for low-power IoT (Internet of Things) devices having limited memory, for example wireless sensor nodes. A typical Contiki configuration is suitable for a microcontroller with 2 kilobytes of RAM and 40 kilobytes of ROM memory [40]. With this operating system is running on connected resource-constrained devices, it is possible to develop many different applications.

3.6.3 Building a PUF Model

The protocol assumes that the verifier has access to the PUF in each device at the registration stage and able to collect sufficient number of CRPs to construct a software model of the PUF, which will be stored in its database. The verifier will also need the fuse the programmable wires to challenge-response interface, so that further access to the PUF through the primary inputs and outputs of the IoT device. In this work, a model of the PUF incorporated in each device is derived in the same manner explained in Sect. 3.6.3 are used.

3.6.4 Functional Verification

A wireless network, which has a server and two clients, was constructed. First, the server successfully verified the respective identities of the two clients, and then the two clients agreed on a secret key using the proposed protocol. In each case, this process was carried out 10 times.

3.6.5 Metrics of Evaluation

The experiment runs mainly in two phases: memory utilisation and energy estimation.

a) **Memory utilisation**: To measure the memory utilisation of each implemented protocols, the arm toolchain command "arm-none-eabi-size" was used [39]. This command gives a breakdown of the data used up by both the RAM and flash memory. After the program code has been compiled and uploaded to the device, the Arm toolchain is then run on the terminal to determine the memory utilisation, as shown in Fig. 3.9.

b) **Energy estimation**: To measure the energy consumption of each-component in the devices, an application called the energest module (for implementation code readers are referred to Appendix A) on the Contiki operating system was used. The energest module measures time by taking the readings of clock ticks while the device is in the receive state, transmit state, processing (CPU) mode and low power mode. The processing time of each component in milliseconds (ms) is calculated by Formula (3.1).

$$\text{Processing time } [ms] = \frac{\text{Energest\_Value (ticks)} \times 1000}{\text{CLOCK\_SECOND}} \tag{3.1}$$

Fig. 3.9 Memory usage in bytes estimated by "arm-none-eabi-size"

In order to measure the energy consumption of these states, the following formula was used:

$$E = \frac{\text{Energest\_Value} \times \text{Current} \times \text{Voltage}}{\text{CLOCK\_SECOND}} \qquad (3.2)$$

where the *Energest_Value* is read off the terminal directly while the program is running, where the *voltage* and *current* at different operating levels are obtained from Table 3.3. This experiment is conducted in order to measure how efficient the proposed protocol is on constrained devices. The results from the proposed protocol are compared with the results of a DTLS-based protocol in the following section. For the PSK-DTLS protocol, measurements are limited to the handshake phase, since it is the part which is concerned with device authentication.

Based on all of the above-mentioned metrics, the following section will evaluate the memory and energy-related cost of the implementation of the proposed protocol.

3.7 Evaluation and Cost Analysis

In this section, the PSK-DTLS, the proposed protocol and the UDP connection (i.e. without security) have been evaluated by measuring energy consumption and memory utilisation in resource-constrained devices (i.e. the Zolertia Zoul RE-mote). It must be noted that the experimental results presented for the client correspond to the measurements on device *A*.

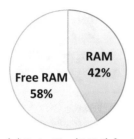 

(a) Data Size(RAM) footprint   (b) Code size (ROM) footprint

in Byte	.text	.data	.bss	ROM	RAM
Client(zoul)	55170	832	12799	56002	13631
Server(zoul)	59505	1664	12993	61169	14657

Fig. 3.10 Memory utilisation of the proposed protocol on Zoul

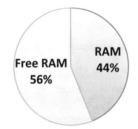 

(c) Data Size(RAM) footprint   (d) Code size (ROM) footprint

in Byte	.text	.data	.bss	ROM	RAM
Client(zoul)	64904	750	13623	65654	14373
Server(zoul)	67364	1500	13733	68864	15233

Fig. 3.11 Memory utilisation of the PSK-DTLS implementation on Zoul

3.7.1 Estimation of Memory Usage

Figures 3.10, 3.11 and 3.12 represent the total ROM and RAM usage of the three protocols, along with the usage details of the memory sections: namely *bss, data* and *text*. As described in [39], the RAM usage in the system is the sum of the *bss* and *data* sections. The *bss* area includes dynamic variables, whereas the *data* area composes static variables. The ROM usage in the system is the sum of the *text* and *data* sections. The ROM includes other sections of the program, which are static during the run time of the program.

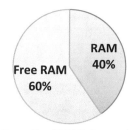

(e) Data Size(RAM) footprint   (f) Code size (ROM) footprint

in Byte	.text	.data	.bss	ROM	RAM
Client(zoul)	50022	479	12764	50501	13243
Server(zoul)	51234	958	12878	52192	13836

Fig. 3.12 Memory utilisation of UDP implementation without security on Zoul

Table 3.4 Percentages of memory utilisation of three implementations on Zoul

	Available (Kbytes)	PSK-DTLS Usage (%)	Proposed	UDP
ROM (flash memory)	512	**13.42**	**11.01**	**9.63**
RAM	32	**43.86**	**42.3**	**40.41**

The results obtained from the proposed protocol implementation reveals that on the client 13.631 bytes and 56.002 bytes of RAM and ROM were used, respectively, which accounts for 42.3% of the total available RAM and 11.01% of the total available ROM. When comparing the aforementioned protocol with the PSK-DTLS implementation on the Zoul RE-mote, it can be seen that the proposed protocol performs slightly better than the PSK-DTLS implementation, which itself used 65.654 bytes of ROM and 14.373 bytes of RAM memory. When the program is implemented without security on the client, the memory usage stood at 13.243 bytes and 50.501 bytes, respectively, which accounts for 40.41% of the available RAM and 9.63% of the available ROM.

3.7.2 Discussion of Memory Utilisation Results

The above results for memory utilisation indicate that the overhead resulting from the use of the proposed protocol would not result in a massive strain on the resource-constrained device. In fact, the AASURE protocol requires an additional 2% memory resources compared to UDP, as shown in Table 3.4.

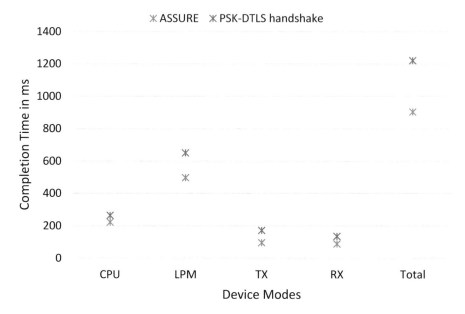

Fig. 3.13 Average completion time of DTLS handshake and the proposed protocol for one authentication cycle

3.7.3 Estimation of Completion Time

Figure 3.13 shows the procession times corresponding to different tasks associated with each protocol. These include the periods during which the device's processor is active (CPU), on low power mode (LPM), and transmitting (Tx) and receiving (Rx), data. These results, calculated based on Eq. (3.1), indicate that the ASSURE protocol needs less than one second on average to complete compared to an average of 1.2 second required by the PSK-DTLS protocol.

3.7.4 Estimation of Energy Consumption

A linear method is used when measuring online energy consumption for the devices used in this experiment. The total energy (E) consumption of all components is computed as:

$$E = (I_m t_m + I_l t_l + I_t t_t + I_r t_r) \times V \tag{3.3}$$

This formula has the following variables: V stands for supply voltage; I_m for the current draw of the microprocessor during operation; t_m for operation time of the

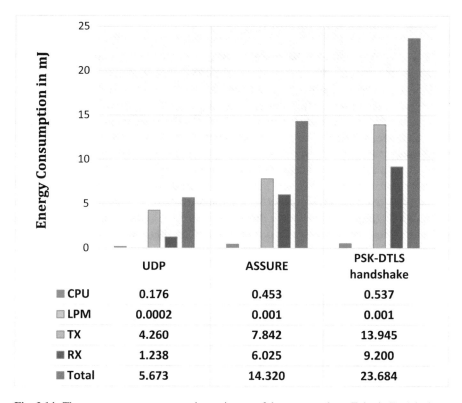

	UDP	ASSURE	PSK-DTLS handshake
■ CPU	0.176	0.453	0.537
▨ LPM	0.0002	0.001	0.001
▨ TX	4.260	7.842	13.945
■ RX	1.238	6.025	9.200
■ Total	5.673	14.320	23.684

Fig. 3.14 The average energy consumption estimates of three protocols on Zolertia Zoul device

microprocessor; I_l and t_l for current draw and time in low power mode; I_t and t_t for current draw and communication time in transfer mode, that is Tx; I_r and t_r for current draw and communication time in receive mode, that is Rx.

In order to evaluate the energy consumption, each protocol considered was run 10 times. Energy measurements were taken from each protocol cycle based on the power breakdown of the Zolertia Zoul node, detailed in Table 3.3. Then the average of the energy consumption results is computed accordingly for each protocol. The results in Fig. 3.14 show that the ASSURE protocol requires approximately 60% of the energy required by the PSK-DSTL. This amount in saving is significant for energy-constrained devices. It can also be observed that the CPU mode and LPM mode of the device consume very little energy compared with the receive and transfer modes, which consume the most energy. This is because the device is almost always on the listen and transfer modes while expecting connections. This observation is going to be used in Chap. 4 to develop an energy-efficient authentication scheme by reducing the number of required transaction between a client and a server.

3.8   Conclusion

This chapter has outlined the principle of a hardware-based authentication and
peer-to-peer key agreement protocol for resource-constrained devices. The solution
combines the use of physically unclonable function technology and light symmetric
cipher to achieve energy efficiency and security against a range of security attacks.
To validate the functionality, a network of three nodes (two clients and one server)
has been created using Zolertia Zoul RE-mote devices. The ARM GCC toolchain
(arm-none-eabi) application has been used to determine memory usage, and the
energest application has been used to monitor energy consumption. According
to this evaluation, the proposed protocol has used less memory and consumed
less energy than the DTLS handshake implementation. Formal security analysis
using the Scyther tool has shown that the proposed solution is resistant to MitM,
eavesdropping, replay, server and client impersonation, de-synchronization, and
PUF model-building attacks.

A.   Appendix A

Energest Module

```
// Project : Energest values printing
// Program name : PrintEnergest.c
// Author : yy6e14
// Date created : 20/7/2019
// Purpose : To print processing time of each components on the device.
    static unsigned long convto_milliseconds(uint64_t time)
{
 return (unsigned long)(time*1000 / RTIMER_ARCH_SECOND);
}
void printEnergest()
{
 energest_flush();

 printf("\nEnergest-values:\n");
 printf(" CPU_mode %lums LPM_mode %lums \n",
 convto _milliseconds(energest_type_time(ENERGEST_TYPE_CPU)),
 convto _milliseconds(energest_type_time(ENERGEST_TYPE_LPM)));
 printf(" Radio LISTEN_mode %lums TRANSMIT_mode %lums \n",
 convto _milliseconds(energest_type_time(ENERGEST_TYPE_LISTEN)),
 convto _milliseconds(energest_type_time(ENERGEST_TYPE_TRANSMIT)));


 unsigned long total_time= to_milliseconds(energest_type_time
 (ENERGEST_TYPE_CPU))
 + convto _milliseconds(energest_type_time(ENERGEST_TYPE_LPM))
 + convto _milliseconds(energest_type_time(ENERGEST_TYPE_LISTEN))
 + convto _milliseconds(energest_type_time(ENERGEST_TYPE_TRANSMIT));
 printf("Completion time: %lu\n", total_time);
 printf("RTIMER_ARCH_SECOND= %lu",RTIMER_ARCH_SECOND);
}
```

References

1. S. Raza, H. Shafagh, K. Hewage, R. Hummen, T. Voigt, Lithe: Lightweight secure CoAP for the internet of things. IEEE Sensors J. **13**(10), 3711–3720 (Oct. 2013). https://doi.org/10.1109/JSEN.2013.2277656
2. G. Arfaoui, X. Bultel, P. Fouque, A. Nedelcu, C. Onete, The privacy of the TLS 1.3 protocol, in *Cryptology ePrint Archive*, Report 2019/749, (2019), pp. 190–210. https://doi.org/10.2478/popets-2019-0065
3. A. Capossele, V. Cervo, G. De Cicco, C. Petrioli, Security as a CoAP resource: An optimized DTLS implementation for the IoT, in *IEEE International Conference on Communications*, vol. 2015-September, (2015, June), pp. 549–554. https://doi.org/10.1109/ICC.2015.7248379
4. G. Lessa dos Santos, V.T. Guimaraes, G. da Cunha Rodrigues, L.Z. Granville, L.M.R. Tarouco, A DTLS-based security architecture for the Internet of Things, in *2015 IEEE Symposium on Computers and Communication (ISCC)*, vol. 2016-February, (2015, July), pp. 809–815. https://doi.org/10.1109/ISCC.2015.7405613
5. D. Mukhopadhyay, PUFs as promising tools for security in internet of things. IEEE Des. Test **33**(3), 103–115 (2016, June). https://doi.org/10.1109/MDAT.2016.2544845
6. J. Delvaux, R. Peeters, D. Gu, I. Verbauwhede, A survey on lightweight entity authentication with strong PUFs. ACM Comput. Surv. **48**(2), 1–42 (Oct. 2015). https://doi.org/10.1145/2818186
7. A.C.D. Resende, K. Mochetti, D.F. Aranha, *Lightweight Cryptography for Security and Privacy*, vol 9542 (Springer, Cham, 2016)
8. K.B. Frikken, M. Blanton, M.J. Atallah, Robust authentication using physically unclonable functions, in *Lecture Notes in Computer Science*, vol. 5735 LNCS, (2009), pp. 262–277
9. Ü. Kocabaş, A. Peter, S. Katzenbeisser, A.-R. Sadeghi, Converse PUF-based authentication, in *Lecture Notes in Computer Science*, vol. 7344 LNCS, (2012), pp. 142–158
10. B. Halak, *Physically Unclonable Functions From Basic Design Principles to Advanced Hardware Security Applications*, 1st edn. (Springer, Cham, 2018)
11. Y. Gao, H. Ma, S.F. Al-Sarawi, D. Abbott, D.C. Ranasinghe, PUF-FSM: A controlled strong PUF. IEEE Trans. Comput. Des. Integr. Circuits Syst. **37**(5), 1–1 (2017). https://doi.org/10.1109/TCAD.2017.2740297
12. T. Kothmayr, C. Schmitt, W. Hu, M. Brünig, G. Carle, DTLS based security and two-way authentication for the Internet of Things. Ad Hoc Netw. **11**(8), 2710–2723 (2013, November). https://doi.org/10.1016/j.adhoc.2013.05.003
13. P. Wouters, E.H. Tschofenig, J. Gilmore, S. Weiler, T. Kivinen, Using raw public keys in Transport Layer Security (TLS) and Datagram Transport Layer Security (DTLS). RFC 5741, 1–18 (2014) doi: 2070-1721
14. M. Bafandehkar, S.M. Yasin, R. Mahmod, Z.M. Hanapi, Comparison of ECC and RSA algorithm in resource constrained devices, in *2013 International Conference on IT Convergence and Security (ICITCS)*, (2013, December), pp. 1–3. https://doi.org/10.1109/ICITCS.2013.6717816
15. C.S. Park, W.S. Park, A group-oriented DTLS handshake for secure IoT applications. IEEE Trans. Autom. Sci. Eng. **15**(4), 1920–1929 (2018, October). https://doi.org/10.1109/TASE.2018.2855640
16. S. Raza, L. Seitz, D. Sitenkov, G. Selander, S3K: Scalable security with symmetric keys—DTLS key establishment for the internet of things. IEEE Trans. Autom. Sci. Eng. **13**(3), 1270–1280 (2016, July). https://doi.org/10.1109/TASE.2015.2511301
17. J. Granjal, E. Monteiro, J.S. Silva, On the effectiveness of end-to-end security for internet-integrated sensing applications, in *2012 IEEE International Conference on Green Computing and Communications*, (2012, November), pp. 87–93. https://doi.org/10.1109/GreenCom.2012.23

18. J. Granjal, E. Monteiro, J.S. Silva, End-to-end transport-layer security for Internet-integrated sensing applications with mutual and delegated ECC public-key authentication, in *2013 IFIP Networking Conference, IFIP Networking 2013*, (2013), pp. 1–9
19. Z. Shelby, K. Hartke, C. Bormann, The Constrained Application Protocol (CoAP). RFC7252, 3 (2014, June). https://doi.org/10.17487/rfc7252
20. Tinydtls URL: https://projects.eclipse.org/proposals/tinydtls
21. J. Lee, K. Kapitanova, S.H. Son, The price of security in wireless sensor networks. Comput. Netw. **54**(17), 2967–2978 (2010, December). https://doi.org/10.1016/j.comnet.2010.05.011
22. R. Maes, *Physically Unclonable Functions Constructions, Properties and Applications* (Springer, Berlin/Heidelberg, 2013)
23. M. Burrows, M. Abadi, R. Needham, A logic of authentication. ACM Trans. Comput. Syst. **8**(1), 18–36 (1990, February). https://doi.org/10.1145/77648.77649
24. C. Cremers, S. Mauw, *Operational semantics and verification of security protocols* (Springer, Berlin/Heidelberg, 2012, November)
25. C.J.F. Cremers, P. Lafourcade, P. Nadeau, Comparing state spaces in automatic security protocol analysis. Lect. Notes Comput. Sci. (including Subser. Lect. Notes Artif. Intell. Lect. Notes Bioinformatics) **5458**, 74–94 (2009). https://doi.org/10.1007/978-3-642-02002-5-5
26. C.J.F. Cremers, The Scyther tool: Verification, falsification, and analysis of security protocols, in *Computer Aided Verification*, vol. 5123 LNCS, (Springer, Berlin/Heidelberg, 2008), pp. 414–418
27. R. Patel, B. Borisaniya, A. Patel, D. Patel, M. Rajarajan, A. Zisman, Comparative analysis of formal model checking tools for security protocol verification, in *Communications in computer and information science*, vol. 89 CCIS, (2010), pp. 152–163
28. A. Armando et al., The AVISPA tool for the automated validation of internet security protocols and applications, in *Computer Aided Verification*, vol. 3576, (2005), pp. 281–285
29. B. Blanchet, An efficient cryptographic protocol verifier based on prolog rules, in *Proceedings. 14th IEEE Computer Security Foundations Workshop, 2001*, vol. 96, (2005), pp. 82–96. https://doi.org/10.1109/CSFW.2001.930138
30. G.E. Suh, S. Devadas, Physical unclonable functions for device authentication and secret key generation, in *2007 44th ACM/IEEE Design Automation Conference*, vol. 129, (2007, June), pp. 9–14. https://doi.org/10.1109/DAC.2007.375043
31. L. Daihyun, J.W. Lee, B. Gassend, G.E. Suh, M. van Dijk, S. Devadas, Extracting secret keys from integrated circuits. IEEE Trans. Very Large Scale Integr. Syst. **13**(10), 1200–1205 (2005, October). https://doi.org/10.1109/TVLSI.2005.859470
32. U. Ruhrmair et al., PUF Modeling attacks on simulated and silicon data. IEEE Trans. Inf. Forensics Secur. **8**(11), 1876–1891 (2013, November). https://doi.org/10.1109/TIFS.2013.2279798
33. M. Barbareschi, P. Bagnasco, A. Mazzeo, Authenticating IoT devices with physically unclonable functions models, in *2015 10th International Conference on P2P, Parallel, Grid, Cloud and Internet Computing (3PGCIC)*, vol. 2015, November, pp. 563–567. https://doi.org/10.1109/3PGCIC.2015.117
34. W. Che, PUF-based authentication invited paper, in *IEEE/ACM international conference on Computer-aided design*, (2015), pp. 337–344
35. A. Perrig, R. Szewczyk, V. Wen, D. Culler, J.D. Tygar, SPINS: Security protocols for sensor networks, in *Proceedings of the 7th annual international conference on Mobile computing and networking - MobiCom'01*, (2001), pp. 189–199. https://doi.org/10.1145/381677.381696
36. M.S. Mispan, B. Halak, Z. Chen, M. Zwolinski, TCO-PUF: A subthreshold physical unclonable function, in *2015 11th Conference on Ph.D. Research in Microelectronics and Electronics (PRIME)*, (2015, June), pp. 105–108. https://doi.org/10.1109/PRIME.2015.7251345
37. M.S. Mispan, B. Halak, M. Zwolinski, Lightweight obfuscation techniques for modeling attacks resistant PUFs, in *2017 IEEE 2nd International Verification and Security Workshop (IVSW)*, (2017, July), pp. 19–24. https://doi.org/10.1109/IVSW.2017.8031539
38. "Zolertia RE-Mote Revision B," 2016.

39. A. Velinov, A. Mileva, Running and testing applications for Contiki OS using Cooja simulator, in *International Conference on Information Technology and Development of Education*, (2016), pp. 279–285
40. M. Sethi, J. Arkko, A. Keranen, End-to-end security for sleepy smart object networks, in *Proceedings – Conference on Local Computer Networks, LCN*, (2012, October), pp. 964–972. https://doi.org/10.1109/LCNW.2012.6424089

Chapter 4
TIGHTEN: A Two-Flight Mutual Authentication Protocol for Energy-Constrained Devices

Yildiran Yilmaz and Basel Halak

Abstract This chapter discusses the design, implementation and evaluation of a two-flight authentication protocol. The latter combines the use of a lightweight symmetric cipher (RC5) with elliptic curve cryptography (ECC) to reduce the number of required interactions between the prover and the verifier, hence reducing the amount of dissipated energy associated with communication. The security of the proposed protocol is formally verified using Scyther. Additionally, a wireless network is constructed using Zolertia Re-mote IoT devices, for functional verification and cost analysis. The results indicated that the proposed protocol can achieve up to 57% energy saving compared to existing solutions such as those based on raw public key (RPK)–based DTLS handshake.

Keywords PUF · Authentication protocol · RC5 · Elliptic curve cryptography (ECC) · Communication energy · Security · Scyther · IoT · DTLS

4.1 Introduction

Energy efficiency is an important design factor for security protocols for resource-constrained devices. There are a number of solutions that have been proposed to reduce energy costs of authentication schemes by reducing computation complexity, such as [1–5]. However, these protocols still require at least four or more (e.g. RPK-DTLS has six) interactions between the client and the server to complete mutual authentication. This causes a large amount of energy to be consumed on data transmission and reception actions.

As demonstrated experimentally in Chap. 3, a sizable portion of the energy costs of security protocols is due to communication tasks. This has also been echoed in [2], wherein the authors suggest that the reduction of the number of transaction

Y. Yilmaz · B. Halak (✉)
University of Southampton, Southampton, UK
e-mail: yy6e14@southamptonalumni.ac.uk; basel.halak@soton.ac.uk; bh9@ecs.soton.ac.uk

© Springer Nature Switzerland AG 2021
B. Halak (ed.), *Authentication of Embedded Devices*,
https://doi.org/10.1007/978-3-030-60769-2_4

flights between the client and the server is the most efficient approach to reduce the overall energy requirements. Therefore, this chapter combines the use of a lightweight symmetric cipher (RC5) with elliptic curve cryptography (ECC) to develop a mutual authentication protocol that requires only two interactions between a prover and a verifier. This will be referred to as TIGHTEN (Two-flIGHts mutual auThentication for Energy constraiNed devices). Formal security analysis using Scyther demonstrates that the protocol is resilient to de-synchronisation, eavesdropping, man-in-the-middle, client impersonation and server spoofing attacks. The functionality of the proposed system is verified by constructing a two-node wireless network using Zolertia Re-mote IoT devices. The energy costs and memory usage of the proposed security mechanism are estimated and compared with existing solutions, including tinyDTLS (a raw public key (RPK) DTLS) and UDP-only-based communication implemented on the same network. Results show that TIGHTEN saves up to 57% of total energy consumption per authentication cycle compared to tinyDTLS.

4.2   Chapter Overview

The organisation of this chapter is as follows. Section 4.3 provides a primer on elliptic curve cryptography. Sections 4.4 and 4.5 present the proposed protocol and related security analysis, respectively. Section 4.6 defines the experimental analysis method. Section 4.7 evaluates the memory and energy-related costs and the security capabilities of the proposed protocol. Finally, Section 4.8 concludes the chapter.

4.3   A Primer on Elliptic Curve Cryptography (ECC)

Elliptic curve cryptography is an asymmetric cryptosystem [6] which offer the same level of security compared to RSA while needing shorter keys [7]; in fact NIST-recommended RSA key lengths is at least 2048 bits compared to 224 for ECC keys. Additionally, the required memory resources and energy-related costs of its arithmetic operations are smaller than those associated with RSA cryptosystems [7]. ECC systems can be constructed over prime fields F_p or binary fields F_{2^m} [8]. The former is adopted in this work.

4.3.1   Definition

An elliptic curve over real numbers may be defined as the set of points (x, y) which satisfy an elliptic curve equation below. It also includes a "point at infinity:"

$$y^2 = x^3 - 1 \qquad y^2 = x^3 + 1 \qquad y^2 = x^3 - 4x \qquad y^2 = x^3 - x$$

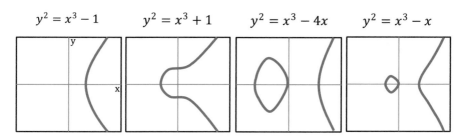

Fig. 4.1 Some example of the elliptic curves

$$y^2 = x^3 + ax + b$$

where

A and b are real numbers such that

$$4a^3 + 27b^2 \neq 0$$

This condition ensures the elliptic curve equation has three distinct roots. Otherwise, geometrically speaking, the curve can intersect with itself, in which case the elliptic curve cannot form a group. Each choice of the numbers a and b yields a different elliptic curve, as shown in Fig. 4.1.

4.3.2   Elliptic-Curve-Based Group Operations

One of the main reasons for the importance of elliptic curves is that they have a natural Abelian group structure. To understand the group operations, it is useful to start with elliptic curves over the real field $F = R$.

4.3.2.1   Point Addition

Point addition is the process of creating another point in the form of $P(x_p, y_p)$ by adding two different points on the elliptic curve, such as $K(x_K, y_K)$ and $L(x_L, y_L)$.

Let us first assume that $y_K \neq -y_L$

To add the points K and L, a line is drawn through them, which intersects the curve at a third point $(-P)$, which is the reflection point P relative to the x-axis. By definition, $K + L = P$. The process is shown geometrically in Fig. 4.2.

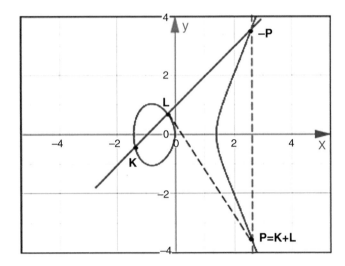

Fig. 4.2 ECC geometrical point addition, example 1

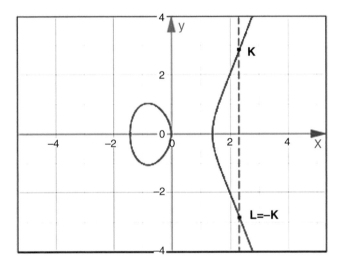

Fig. 4.3 ECC geometrical point addition example 2

If $y_K = -y_L$, that is the points K and L are symmetrical with respect to the x-axis, in this case, the points K and L intersect the curve at the infinite point 'O', as shown in Fig. 4.3.

Thus,

$$K + L = O$$

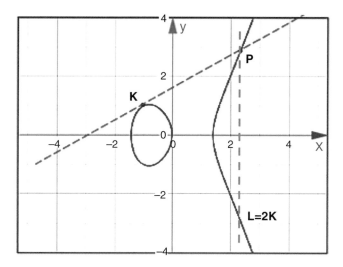

Fig. 4.4 ECC geometrical point addition example 3

where O is the "point of infinity".

4.3.2.2 Point Doubling

Point doubling is the process adding the point to itself. Let us first assume that

$$y_K \neq 0$$

In this case, to find $2\,K$, a line tangent to the curve is drawn through K, which intersects the curve on point $-P$, which is the reflection of point P relative to the x-axis. By definition, $2\,K = P$. The process is shown geometrically in Fig. 4.4.

The sum of K added to itself gives the point L by mirroring the intersection point along x-axis. The process of $K + K = 2\,K$ is shown geometrically in Fig. 4.4.

If $y_K = 0$, since the tangent of the curve at point K is parallel to the y-axis, in this case, this tangent intersects the curve at the infinite point 'O', as shown in Fig. 4.5 thus $K + K = O$.

4.3.2.3 Scalar Point Multiplication

This is achieved through repeated point. For example, if $k = 5$, the point of $R = kQ$ can be found as follows, using two doubling operation and one point addition as follows.

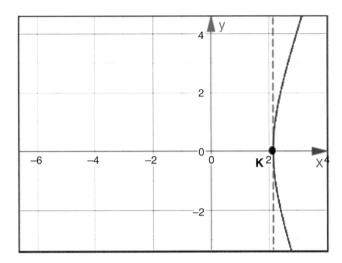

Fig. 4.5 ECC geometrical point addition example 4

$$Q + Q = P$$
$$P + P = L$$
$$R = L + Q$$

Similarly, we define the elliptic curve E over prime fields Z_P as follows:

$$E = E(F) = \{(x, y) \mid x, y \in F, y^2 = x^3 + ax + b \,(\text{mod } p)\} \cup \{\infty, \infty\}$$

such that

$$4a^3 + 27b^2 \neq 0 \,(\text{mod } P)$$

P is a large prime number.

Point addition and doubling calculation are similar; however, in this case, geometrical operations are not possible but corresponding equations remain the same, with the crucial difference that all mathematical operations are performed mod P.

4.3.2.4 Elliptic Curve Discrete Logarithm Problem

The security of elliptic curve cryptography depends on how difficult it is to determine i given iP and P. This is referred to as the elliptic curve discreet logarithm problem (ECDLP). One of the fastest known techniques to solve ECDLP is called

Pollard rho method, which require $O\left(\sqrt{n}\right)$, where n is the size of the underlying field.

Compared to factoring integers or polynomials, one can use much smaller numbers for equivalent levels of security.

4.3.2.5   Elliptic Curve–Based Diffie-Hellman Scheme

The elliptic curve–based Diffie Hellman (ECDH) is a key exchange algorithm [8] that allows two communicating parties to create a shared cryptographic key that is used for symmetric key cryptosystems. Both parties must first share the field parameters, including:

- E is a third-degree polynomial elliptic curve equation which defines the group, which includes that a and b are constants.
- p is a large prime.
- n is the number of elements in the group.
- G is generator point, that is all other points in the ECC group can be generated by multiplying G with an integer i.

The detailed steps of the ECDH protocol are shown in Fig. 4.6, wherein the communication is assumed to take place between a server (S) and device (D):

4.3.2.6   Elliptic Curve-Based Digital Signature Algorithm

The elliptic curve–based digital signature algorithm (ECDSA) is a digital signature scheme used to provide message integrity and non-repudiation [8, 9]. In this case, communicating parties should first share the field parameters as above. ECDSA consists of two main algorithms, namely signature generation algorithm and signature verification algorithm, as described in the following two sections (Fig. 4.7).

Digital Signature Generation

When signing a message m, device D uses domain parameters and private key (P_D) information to perform the following:

(i) Calculate e = Hash (m); the result of the hash value is converted to an integer e.
(ii) Generate a random integer number $r \leftarrow \{1, \ldots, n-1\}$
(iii) Calculate $\alpha = x_1 \bmod n$, where $(x_1, y_1) = rG$. If $\alpha = 0$, then return to (ii), otherwise continue.
(iv) Calculate $\beta = r^{-1}(e + s'\alpha) \bmod n$. If $\beta = 0$, then return to (ii), otherwise continue. Signature pairs are (α, β).

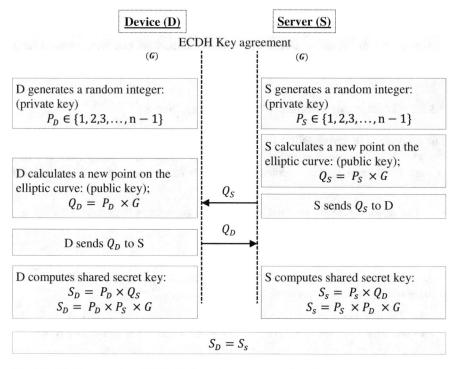

Fig. 4.6 Elliptic curve–based Diffie-Hellman key agreement algorithm

At the end of the above steps, device D will have created a digital signature pair (α, β) for the message m. These are sent to the recipient server S for verification.

Digital Signature Verification

In order for the server S to verify the digital signature of a message m received from a device D. The server will need to use the domain parameters $FP = (a, b, G, p, h, n)$ and the public key information of the device, and perform the following operations:

(i) If (α, β) are integer numbers withing the range $\{1, n - 1\}$, the signature is valid, otherwise invalid, then the verification is terminated.

(ii) From the incoming message m, the hash value with $e = \text{Hash}(m)$ is recalculated and the resulting bit index is converted to e integer value.

(iii) Calculate $w = \beta^{-1} \bmod n$, then $u_1 = ew \bmod n$ and $u_2 = \alpha w \bmod n$, then $(x_1, y_1) = u_1 G + u_2 S'$.

(iv) If $x_1 = \alpha \bmod n$, then the signature is authenticated; otherwise the server terminates the verification.

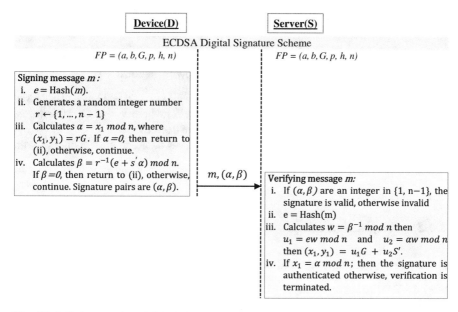

Fig. 4.7 Elliptic curve–based digital signature algorithm

Table 4.1 TIGHTEN protocol notations

	Description
K_i	The symmetric key for RC5 exchanged by ECDH
d, D	Private, public keys of a device used in ECDH to generate symmetric key K_i
s, S	The private, public key (also admitted as a message when signing S in ECDSA) of the server used in ECDH to generate symmetric key K_i
(α, β)	Signature pair of S (the public key of the server)
s', S'	Private, public keys of the server used in ECDSA.
ID_i	The ID of the device i
r	Random number used when S (the public key of the server) signing in ECDSA.

4.4 TIGHTEN Protocol Description

The proposed protocol is based on the use for Elliptic Curve Diffie-Hellman (ECDH) for key exchange and the use of RC5 to encrypt the communicated data. Message integrity is also achieved using the ECDSA scheme. The proposed protocol consists of two stages, registration and verification. Table 4.1 summarises the notations used to describe the protocol.

4.4.1 Registration Stage

In the registration stage, the server and client devices agree on the elliptic curve domain parameters as above and perform the following actions:

1. The server generates an integer s′ randomly $s' \leftarrow \{1, \ldots, n-1\}$ as a private key regarding ECDSA and generates its public key S', where $S' = s'G$. Afterwards, the public key S' of the server is installed on the devices.
2. Domain parameters and s' are stored in the server database corresponding with the device identifier (ID) and MAC address.

4.4.2 Verification Stage

In the verification stage, the following ten steps, depicted in Fig. 4.8, are carried out:

Step 1. The server generates private key s randomly to use in ECDH, where $s \leftarrow \{1, \ldots, n-1\}$, and calculates public key such that $S = sG$ (S is also regarded as a message for ECDSA).

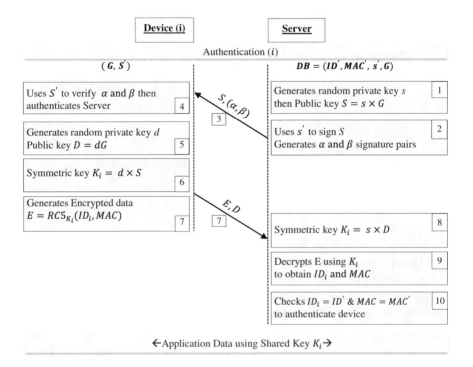

Fig. 4.8 TIGHTEN Description

Step 2. The server signs S through ECDSA such that:

   (i).   e = Hash-One(S)
   (ii). Generate random number $r \leftarrow \{1, \ldots, n-1\}$.
   (iii). Calculate $\alpha = x_1 \bmod n$, where $(x_1, y_1) = rG$. If $\alpha = 0$, then return to (ii), otherwise continue.
   (iv). Calculate $\beta = r^{-1}(e + s'\alpha) \bmod n$. If $\beta = 0$, then return to (ii), otherwise continue. At last, signature pairs are (α, β).

Step 3. The server transmits S and its signature pair (α, β) to the device.
Step 4. When the S and its signatures (α, β) are taken by the device, the verification of S by the device is such that:

   (i).   If (α, β) are an integer in $\{1, n-1\}$, the signature is valid, then continue, otherwise invalid, then the verification is terminated.
   (ii). e = Hash-One(S).
   (iii). Device calculates $w = \beta^{-1} \bmod n$, then $u_1 = ew \bmod n$, and $u_2 = \alpha w \bmod n$, then $(x_1, y_1) = u_1 G + u_2 S'$.
   (iv). If $x_1 = \alpha \bmod n$, then the server is authentic; otherwise, the device terminates the verification.

Step 5. If the server authentication is successful, the device generates private key randomly d in $\{1, n-1\}$ to use in ECDH, then calculates public key such that $D = dG$.
Step 6. The device calculates the secret key $K_i = dS = d(sG) = dsG$.
Step 7. The device encrypts MAC and ID_i by $E = RC5_{K_i}(ID_i, MAC)$. Then the device sends its public key D and encrypted E to the server.
Step 8. When the D and E are received by the server, it calculates the symmetric key such that $K_i = sD = s(dG) = dsG$.
Step 9. To obtain the device ID_i and MAC, the server decrypts E.
Step 10. The server will match the ID_i with ID' in the database and find the MAC corresponding with the device ID. If $ID_i = ID'$ and $MAC = MAC'$, then the device is also authenticated, otherwise rejected.

Note that the ECDH is used to generate the symmetric key K_i on both side to encrypt and decrypt the ID and MAC address with the RC5 algorithm. The ECDSA is used to ensure data integrity and authentication.

4.5   Security Analysis

This section describes the security analysis of the TIGHTEN protocol. First, a system model and an attacker model are characterised to define the security properties provided by the proposed protocol – namely, *mutual authentication, confidentiality, availability, integrity and resistance to the following attacks: man in the middle, eavesdropping, replay, client impersonation, server spoofing, modification and de-*

Fig. 4.9 Wireless sensor device connection scenario

synchronisation. These properties are validated using an automatic tool for the analysis of security protocols named Scyther.

4.5.1 System Model

In the system model depicted in Fig. 4.9. The communicating devices perform the authentication protocol by exchanging messages in accordance with their role in order to prove each other's identity.

For example, in the proposed protocol, the device plays the prover role and the server plays the verifier role. For mutual authentication, only two messages in total are exchanged. As shown in Fig. 4.9, this message exchange may be performed through a wireless channel exposed to third parties or adversaries. Therefore, this raises questions about the attacker role, which will be described in the next section.

4.5.2 Attacker Model

The communication channel depicted in Fig. 4.9 may be accessed and tampered with an adversary who listens to all messages. The adversary could potentially play several roles in different instances of the same authentication protocol. They are assumed to have complete knowledge of the protocol design and the corresponding cryptographic algorithms. Relevant examples of attack strategies include man in the middle (MitM), eavesdropping, replay, client impersonation, server impersonation and de-synchronisation attack. It is worth noting here that the attacker is not assumed to be able to have physical access to the device, hence cannot perform hardware tampering attacks to retrieve stored keys. If this were a realistic threat, then hardware-based protocols, such as that described in Chap. 3, would be more appropriate.

4.5.3   Security Properties

In this section, the security properties of the proposed protocol will be analysed, including the resistance to the known attacks and the security services, as stated in Sect. 4.5.

Mutual Authentication The server is first authenticated by the device in the following manner: through the first flight, the signed public key (S) and the signature pairs (α, β) are verified with the public key (S') of the server used in ECDSA. Through the second flight, the device *IDs* and *MAC* address are transferred in an RC5 encrypted data using a different symmetric key K_i for each authentication cycle. The server can decrypt encrypted data (E) and obtain the *ID* and *MAC*. As the specific identifiers (*ID*) of all devices are recorded in the server database during the registration stage, the server checks if the received *ID* equals to the stored *ID'* and the received *MAC* is equal to the stored *MAC'*, then authenticates the device. Accordingly, the protocol ensures mutual authentication.

De-synchronisation Attack Resistance and Availability The proposed protocol does not require the device's identity and any sensitive data to be updated for subsequent authentications. Therefore, the protocol is not vulnerable to de-synchronisation attacks, and the device and server perform the protocol in a synchronised and available manner.

Eavesdropping and Man-in-the-Middle Attack Resistance Even if an attacker captures messages transferred between the server and the device, it is impossible to obtain critical information from the messages without the private keys (d and s) because the critical data is either signed with the ECDSA or encrypted with the RC5. In messages, a change made by third parties can be detected since S is signed with s' private key, and both *ID* and the *MAC* are sent as encrypted. Then *ID* and *MAC* after being decrypted are compared to the stored *ID'* and *MAC'*. Accordingly, the protocol provides resistance to man-in-the-middle and eavesdropping attacks.

Client Impersonation and Server Spoofing Attack Resistance To impersonate a device, the attacker must generate K_i under the elliptic curve Diffie-Hellman problem (ECDHP) [8] and crack RC5, all of which are impossible to achieve in the duration of an authentication cycle. Since the client ID_i and *MAC* are encrypted with RC5 and the *MAC* and *ID* received by the server have to match with their counterparts previously stored in the server database, client impersonation is not possible. Similarly, the adversary cannot spoof the server, because this requires the secret key K_i that is unreachable without obtaining d, s. Therefore, the protocol is resistant to client impersonation and server spoofing attacks.

Replay Attack Resistance Since C_i, S, α and β are generated randomly, capturing and replaying them to the client does not generate K_i from the past authentication. This is because the device creates randomly a fresh private key to generate the new session key K_i. Likewise, replaying D and E of the previous session cannot cause the server to generate K_i of the previous authentication.

Confidentiality Client *ID* and *MAC* are sent as encrypted text and K_i changes for each session; the attacker cannot obtain any information without calculating K_i or breaking encryption.

Modification Attack Resistance and Integrity Because the ECDSA generates the signature pair (α, β) on the server side, the changes on them can be recognised by the client, and any changes made to *D* will provide the wrong K_i, then inaccurate *ID* and *MAC*. Therefore, the protocol provides integrity and refuses the modifications.

4.5.4 Defining the Proposed Protocol in Scyther

This section shows how of the TIGHTEN protocol can be described using the programming language of Scyther for systematic security analysis, as shown in Listing 4.1. This section also explains the roles, variables, function, events and claims used in this analysis.

Listing 4.1 The TIGHTEN Protocol Model in Scyther Jargon

```
// TIGHTEN protocol description
secret MAC: Function;
hashfunction H1;
//sk(V), pk(V) denotes the private key
//and corresponding public key of V respectively
protocol proposedProtocol(P,V)
{
      role P
      {
              fresh d: Nonce;
              fresh D:Nonce;
              fresh K: Nonce;
              var S;
              var E;
              var alfa;
              var beta;

              recv_1(V, P, S, alfa, beta);
              match(K, k(P,V));
              match(E, {P,MAC}k(K));
              send_2(P, V, E, D);

              claim_P1(P, Secret, alfa);
              claim_P2(P, Secret, beta);
              claim_P3(P, Secret, MAC);
              claim_P4(P, Secret,K);
```

(continued)

```
            claim_P5(P, Nisynch);
            claim_P6(P, Commit, V, alfa, beta,MAC,K,D,S);

     }

     role V
     {
            fresh alfa: Nonce;
            fresh beta: Nonce;
            fresh K: Nonce;
            fresh S:Nonce;
            var D;
            var E;

            match(alfa, (S, {H1(S)}sk(V)));
            match(beta, (S, {H1(S)}sk(V)));
            send_1(V, P, S, alfa, beta);
            match(K, k(P,V));
            recv_2(P, V, E, D);

            claim_V1(V, Secret, alfa);
            claim_V2(V, Secret, beta);
            claim_V3(V, Secret, MAC);
            claim_V4(V, Secret,K);
            claim_V5(V, Nisynch);
            claim_V6(V, Commit, P, alfa, beta,MAC,K,D,S);
     }
}
```

For the purpose of analysis, two roles were defined, the device, that is "role P", and the verifier, that is "role V, as described in the protocol description. To commence the protocol process, role V sends the first message to role P which carries its own public key i.e. S, and signature pairs alfa and beta assigned with match events such that "match(alfa,(S,{H1(S)}sk(V)));" and "match(beta,(S,{H1(S)}sk(V)));". In the Scyther jargon, "send_1(V, P, S, alfa, beta);" represents the first message from verifier (V) to prover (P). This message is received with "recv_1(V, P, S, alfa, beta);" event by the prover role P. In the second message, role P sends its own public key D and the encrypted data E using the symmetric key K to role V. This message is formulated as a send event in role P such that "send_2 (P, V, E, D);" and as a receive event in role V such that "recv_2 (P, V, E, D);"

Finally, the security properties of the proposed protocol are defined by six claim events on both roles P and V. With these claims, Scyther knows what needs to be verified. By defining those claims such that "claim_P1(P, Secret, alfa);", "claim_P2(P, Secret, beta);", "claim_P3(P, Secret, MAC);", "claim_P4(P, Nisynch);", "claim_P5(P, Commit, V,

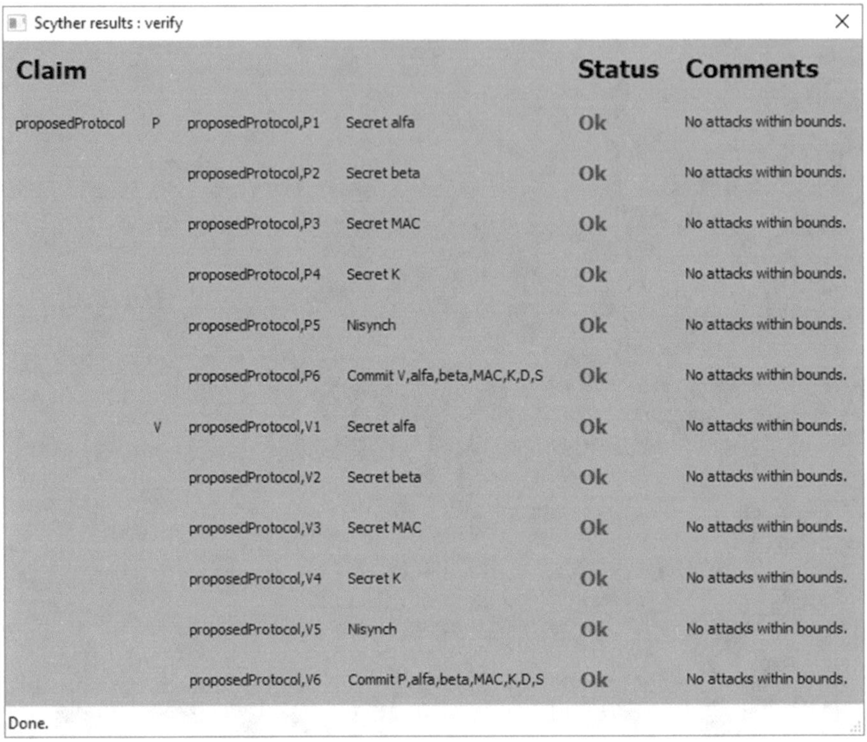

Fig. 4.10 Scyther results

alfa, beta,MAC,K,D,S);", the security of the signature pairs, the MAC value and the symmetric encryption key are checked. This check also includes non-injected synchronisation and data agreement over the set of values, for example "P, alfa, beta,MAC,K,D,S". If Scyther finds a flaw in the protocol, it generates protocol attacks graph, otherwise verifies all claims, as shown in Fig. 4.10.

4.5.5 Proof of Security Properties Using Scyther

The security properties of the TIGHTEN protocol were verified using the Scyther tool [10]. Analysis set-up with Scyther reveals that the TiGHTEN verifies the aforementioned security claims, such as secrecy, nisynch and commit. As defined in [11], the secrecy claims are interpreted as follows: during the realisation of the protocol, transferred data such as key remain secret between both parties, role V and role P. The non-injective and commit claims are interpreted as follows: all events and variables in the definition of the protocol are achieved seamlessly during the implementation of the protocol. Therefore, both parties mutually agree on the content and order of the messages sent and received during the realisation of the protocol. This guarantees de-synchronisation attack resistance and mutual authentication. In addition, if an adversary attacks the communication channel, they cannot obtain any information about the content of the encrypted messages since the symmetric key is unreachable for an adversary as proved by the secret claim. Therefore, decrypting the exchanged messages without the key is impossible. Results for the claims discussed so far guarantee modification attack resistance and Integrity, availability, eavesdropping, man-in-the-middle attack resistance, confidentiality and impersonation attack resistance. Replay attack resistance is ensured as well, since all keys are generated randomly and are changing for each authentication cycle.

Principally, when the Scyther tool finds all the vulnerabilities and attacks, the validation procedure returns results even if the bound is not reached. However, if the verification procedure reaches the bound and finds no flaws, this is reported as "No attacks within bounds", as shown in the output in Fig. 4.10. This confirms that the TIGHTEN protocol satisfies the security properties stated in Sec. 4.5.3.

4.6   Experimental Analysis Method

This section outlines the experimental set-ups and explains the metrics used to evaluate the energy and memory-related costs of the proposed protocol.

4.6.1   The Purpose of the Experiment

The purpose of these experiments is twofold. The first is to verify the functionality of the proposed solution; this is achieved by constructing a wireless network that has a server and a client. The second goal of the experiments is to evaluate the energy consumption and memory utilisation of the proposed scheme compared to existing solutions, namely the RPK-DTLS and UDP-only-based communication protocols. It is worth noting that the following assumptions are made in this work:

- The means of communication is a wireless link.
- The sensor devices are deployed in an environment, restricted to public access, which reduces the risk of device cloning via physical attacks. If the latter were a realistic security threat, then solutions based on PUF technology, such as those discussed in Chap. 3, would be more appropriate
- All the sensor devices deployed in the network are resource-constrained in terms of energy, memory and computation.
- The central device, which is the server, is comparatively a resource-rich device in terms of energy, memory and computation.

4.6.2 Experimental Set-up

The experimental set-up was carried out using Zolertia-Zoul devices [12], the Contiki operating system [13] and a workstation. Contiki OS is set up on VMware with 20 GB hard drive and 4 GB RAM. Two Zoul IoT nodes which have 512 KB ROM and 32 KB RAM used as the resource-limited device, one in the role of server and the other client. The current consumption details of the CPU and radio in active and sleep modes for Zoul device are derived from [12]. The operating voltage is 3.4 V. Those details are used to calculate the energy measurement of specified protocols in Sect. 4.7.3.

Contiki OS was chosen as it was originally designed for resource-limited IoT devices, and it is an open source. A standard Contiki operates typically with 40 KB of ROM and 2 KB of RAM [14].

4.6.3 Metrics of Evaluation

Memory Utilisation A breakdown of the data used in RAM and ROM memory was determined using "msp430-size" module on Contiki OS.

Energy Estimation A breakdown of energy consumed in each component on the device was determined through the use of "energest" module on Contiki. This module estimates processing time counting the ticks in receiving and transmission state, in computation(CPU) and low power mode (LPM). The component energy consumption can be expressed as:

$$(\text{EnergestValue} \times I \times V) / (\text{CLOCKSECOND}) \tag{4.1}$$

where Energest _ Value is taken from the terminal online for each component, and the current(I) and voltage(V) at each state are taken from the device datasheet, and CLOCK _ SECOND is the value of 32768, which donates the number of tick per-second.

Completion Time It is the sum of the processing time required by each component in the system to complete the tasks related to one mutual authentication cycle. This is calculated as follows:

$$\text{Processing Time } [ms] = \frac{\text{Ticks (Energest\_value)} \times 1000}{\text{RTIMER\_ARCH\_SECOND}} \tag{4.2}$$

where:

RTIMER _ ARCH _ SECOND is 32768 for the device used in this experiment.
Ticks stands for the number of ticks calculated using the energest module in Contiki.

4.7   Evaluation and Cost Analysis

This section evaluates three protocols: TIGHTEN, tinyDTLS [4] (based on RPK-DTLS) and UDP-only connection. A two-node wireless network is constructed using the Zolertia Zoul devices. The evaluation metrics discussed above are used to compute the memory utilisation, energy consumption and transaction time required by each protocol to complete one authentication cycle. The latter consists of computation and communication costs. The former includes the resources required to perform the computation associated with the protocol on each node in the system. Communication costs cover two components. First, the transmission of the 256 bits public key and its corresponding signature from the server to the client. Second, the transmission of the client' ID (80 bits), MAC (48 bits) and its public key (256 bits).

4.7.1   Estimation of Memory Usage

Figures 4.11, 4.12 and 4.13, respectively, show the total ROM and RAM require-ments in bytes of the three protocols and their respective percentages compared to available memory resources.

The above results indicate that the memory resources required by the three solutions are comparable. This indicates that the memory usage overheads of implementing secure authentication are relatively small compared with a UDP-only connection (i.e. A non-secure link).

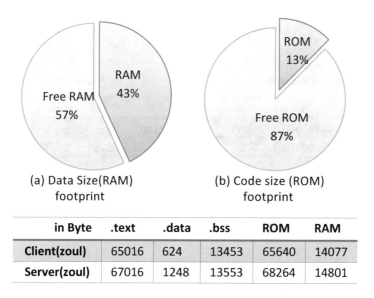

in Byte	.text	.data	.bss	ROM	RAM
Client(zoul)	65016	624	13453	65640	14077
Server(zoul)	67016	1248	13553	68264	14801

Fig. 4.11 Memory footprints of the proposed protocol on Zoul

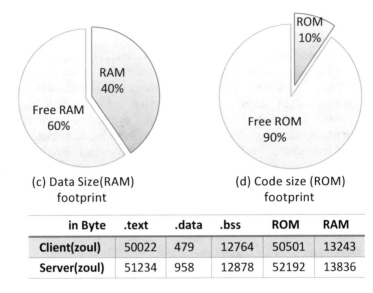

in Byte	.text	.data	.bss	ROM	RAM
Client(zoul)	50022	479	12764	50501	13243
Server(zoul)	51234	958	12878	52192	13836

Fig. 4.12 Memory footprints of UDP without security on Zoul

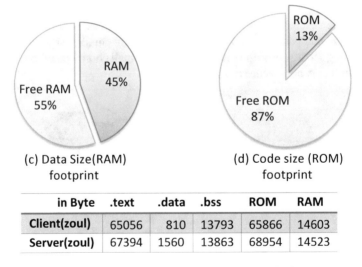

in Byte	.text	.data	.bss	ROM	RAM
Client(zoul)	65056	810	13793	65866	14603
Server(zoul)	67394	1560	13863	68954	14523

Fig. 4.13 Memory footprints of the RPK-DTLS implementation on Zoul

4.7.2 Estimation of Completion Time

Table 4.2 shows the procession times corresponding to different tasks associated with each protocol. These include the periods during which the device's processor is active (CPU), low power mode (LPM), transmitting (Tx) and receiving (Rx) and data, respectively. These figures, calculated based on Eq. (4.2), indicate that the

Table 4.2 Protocol
completion time comparison
results

Type	Timer ticks	Processing time [ms]
TIGHTEN protocol		
CPU	10,557	322
LPM	12,245	373
Transfer(Tx)	2365	72
Receive(Rx)	2234	68
Total	**27,403**	**836**
RPK-DTLS implementation		
CPU	10,721	327
LPM	28,347	865
Transfer(Tx)	6209	189
Receive(Rx)	4898	149
Total	50,177	1531

TIGHTEN protocol requires similar amount of CPU time compared with tinyDTLS but spends a significantly smaller amount of time on data transfer and reception tasks. This was expected, as the proposed solution only needs two flights to complete the authentication cycle compared with six transactions required by tinyDTLS.

4.7.3 Estimation of Energy Consumption

Figure 4.14 shows the energy dissipated by each protocol during one authentication cycle. The results indicate that a significant portion of energy consumption is due to the wireless communication tasks (i.e. Tx and Rx). Another important observation is the significant energy savings (up to 57%) achieved by the TIGHTEN protocol compared to tinyDTLS; this is mainly due to the reduction of the number of transaction

4.7.4 Comparison with Other Protocols

Table 4.3 compares the proposed protocol with authentication protocols reported in [1], [3], [4] and [5] with respect to security properties and the number of required interactions. The solution presented in [1] is DTLS variant, which relies on a trusted third party called IoTSSP (Internet of Things Security Support Provider). The latter is responsible for managing certificates of client nodes and secret keys, authentication of devices, and establishment of secure session among nodes.

The solution in [3] aims to reduce energy consumption by using a reliable gateway to reuse previous sessions. However, in this case, if an eavesdropping adversary manages to obtain packets, they can potentially wage a replay attack

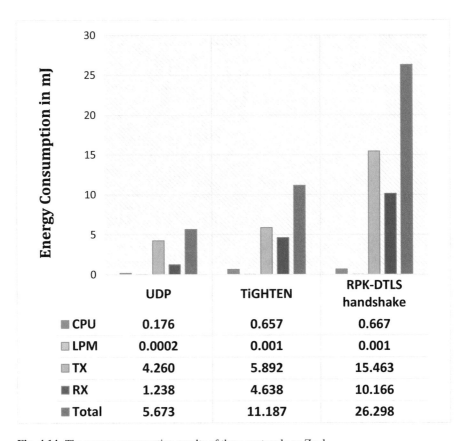

	UDP	TiGHTEN	RPK-DTLS handshake
■ CPU	0.176	0.657	0.667
▢ LPM	0.0002	0.001	0.001
▨ TX	4.260	5.892	15.463
■ RX	1.238	4.638	10.166
■ Total	5.673	11.187	26.298

Fig. 4.14 The energy consumption results of three protocols on Zoul

The tinyDTLS protocol [4], implemented in this work for the comparison, has a security vulnerability, as highlighted [15], because its handshake process has a de-synchronisation risk for communication session if an attacker blocks the last flight during the execution of the protocol. In this last flight, the authenticator (i.e. server) may ask to change the cipher suite that was negotiated previously. However, the resource-constrained node (client) would not be made aware of this change if the last flight was blocked.

The solution in [5] provides authentication and integrity checking. It is based on the use of physically unclonable function technology, which makes it more resilient to device cloning attacks compared to TIGHTEN. However, it needs more transaction per authentication cycle, which makes it less energy efficient.

The results from Table 4.3 indicate that the proposed solution has comparable security properties with existing solutions but provides a more energy-efficient alternative due to the reduction of the number of interaction required to complete a mutual authentication cycle.

Table 4.3 Comparing the proposed protocol with other protocols

	Protocols	[1]	[3]	[4]	[5]	TIGHTEN
	Number of interactions	6	6	6	4	2
Security Capabilities	Mutual authentication	●	○	●	●	●
	Integrity	●	●	●	●	●
	Confidentiality	●	◘	●	●	●
	Replay attacks	–	○	●	●	●
	Eavesdropping	–	●	●	●	●
	De-synchronisation	◘	◘	◘	●	●
	Availability	●	◘	◘	●	●
	Server spoofing	–	–	●	●	●
	Client impersonation	–	●	●	●	●
	Man in the middle	●	○	●	●	●
	Modification attack	–	●	●	●	●

●: Satisfied, ◘: Partly satisfied, ○: Not Satisfied, −: Not provided

4.8   Conclusion

This chapter has developed and implemented a mutual authentication protocol, which requires only two interactions between prover and verifier. It is based on the use of elliptic curve cryptography and the Rivest cipher (RC5). The energy and memory-resource costs of the proposed protocol have been analysed and compared with the RPK-DTLS handshake. The security of the proposed protocol was analysed under different attack scenarios. Cost analysis results show that the proposed solution saves up to 57% energy per authentication cycle compared to tinyDTLS, while needing comparable amount of memory resources. Formal security analysis using the Scyther tool has shown that the proposed solution is resistant to man-in-the-middle, eavesdropping, replay, client impersonation, server spoofing, modification and de-synchronisation attacks.

References

1. G.L. dos Santos, V.T. Guimaraes, G. da Cunha Rodrigues, L.Z. Granville, L.M.R. Tarouco, A DTLS-based security architecture for the Internet of Things, in *2015 IEEE Symposium on Computers and Communication (ISCC)*, vol. 2016-Febru, (2015, July), pp. 809–815. https://doi.org/10.1109/ISCC.2015.7405613
2. T. Kothmayr, C. Schmitt, W. Hu, M. Brunig, G. Carle, A DTLS based end-to-end security architecture for the Internet of Things with two-way authentication, in *37th Annual IEEE Conference on Local Computer Networks – Workshops*, (2012, October), pp. 956–963. https://doi.org/10.1109/LCNW.2012.6424088.
3. F. Van Den Abeele, T. Vandewinckele, J. Hoebeke, I. Moerman, P. Demeester, Secure communication in IP-based wireless sensor networks via a trusted gateway, in *2015 IEEE 10th International Conference on Intelligent Sensors, Sensor Networks and Information Processing, ISSNIP 2015*, (2015, April), pp. 7–9. https://doi.org/10.1109/ISSNIP.2015.7106963

4. "Tinydtls."
5. M. Hossain, S. Noor, R. Hasan, HSC-IoT: A hardware and software co-verification based authentication scheme for Internet of Things, in *2017 5th IEEE International Conference on Mobile Cloud Computing, Services, and Engineering (MobileCloud)*, (2017, April), pp. 109–116. https://doi.org/10.1109/MobileCloud.2017.35
6. U. Kumar, T. Borgohain, S. Sanyal, Comparative analysis of cryptography library in IoT. Int. J. Comput. Appl. **118**(10), 5–10 (2015, May). https://doi.org/10.5120/20779-3338
7. M. Bafandehkar, S.M. Yasin, R. Mahmod, Z.M. Hanapi, Comparison of ECC and RSA algorithm in resource constrained devices, in *2013 International Conference on IT Convergence and Security (ICITCS)*, (2013, December), pp. 1–3. https://doi.org/10.1109/ICITCS.2013.6717816.
8. D. Hankerson, A. Menezes, S. Vanstone, *Guide to Elliptic Curve Cryptography* (Springer, New York, 2004)
9. M. Suarez-Albela, T.M. Fernandez-Carames, P. Fraga-Lamas, L. Castedo, A practical performance comparison of ECC and RSA for resource-constrained IoT devices, in *2018 Global Internet of Things Summit (GIoTS)*, (2018, June), pp. 1–6. https://doi.org/10.1109/GIOTS.2018.8534575.
10. C.J.F. Cremers, The Scyther tool: Verification, falsification, and analysis of security protocols, in *Computer Aided Verification*, vol. 5123, (Springer, Berlin/Heidelberg, 2008), pp. 414–418
11. C. Cremers, S. Mauw, *Operational Semantics and Verification of Security Protocols* (Springer, Berlin/Heidelberg, 2012, November)
12. "Zolertia RE-Mote Revision B," 2016.
13. A. Dunkels, B. Gronvall, T. Voigt, Contiki – A lightweight and flexible operating system for tiny networked sensors, in *29th Annual IEEE International Conference on Local Computer Networks*, (2004), pp. 455–462. https://doi.org/10.1109/LCN.2004.38
14. A. Velinov, A. Mileva, Running and testing applications for Contiki OS using Cooja simulator, in *International conference on information technology and development of education*, (2016), pp. 279–285
15. Y. Sheffer, R. Holz, P. Saint-Andre, Summarizing known attacks on Transport Layer Security (TLS) and Datagram TLS (DTLS). RCF 7457, 1–13 (2015, February). https://doi.org/10.17487/rfc7457

Part III
Emerging Applications of Hardware-based Authentication

Chapter 5
Securing Hardware Supply Chain Using PUF

Leonardo Aniello, Basel Halak, Peter Chai, Riddhi Dhall, Mircea Mihalea, and Adrian Wilczynski

Abstract The complexity of today's integrated circuit (IC) supply chain, organised in several tiers and including many companies located in different countries, makes it challenging to assess the history and integrity of procured ICs. This enables malicious practices like counterfeiting and insertion of back doors, which are extremely dangerous, especially in supply chains of ICs for industrial control systems used in critical infrastructures, where a country and human lives can be put at risk. This paper aims at mitigating these issues by introducing Anti-BlUFf (Anti-counterfeiting Blockchain- and PUF-based infrastructure), an approach where ICs are uniquely identified and tracked along the chain, across multiple sites, to detect tampering. Our solution is based on consortium blockchain and smart contract technologies, hence it is decentralised, highly available and provides strong guarantees on the integrity of stored data and executed business logic. The unique identification of ICs along the chain is implemented by using physically unclonable functions (PUFs) as tamper-resistant IDs. We first define the threat model of an adversary interested in tampering with ICs along the supply chain, then provide the design of the tracking system that implements the proposed anti-counterfeiting approach. We present a security analysis of the tracking system against the designated threat model and a prototype evaluation to show its technical feasibility and assess its effectiveness in counterfeit mitigation. Finally, we discuss several key practical aspects concerning our solution and its integration with real IC supply chains.

Keywords IC supply chain · Counterfeiting · Critical infrastructures · Blockchain · PUF · Tamper detection · FPGA · Security · Tracking

L. Aniello · B. Halak (✉) · P. Chai · R. Dhall · M. Mihalea · A. Wilczynski
School of Electronics and Computer Science, University of Southampton, Southampton, UK
e-mail: l.aniello@soton.ac.uk; basel.halak@soton.ac.uk; xc5g15@soton.ac.uk;
rd12g15@soton.ac.uk; mm8g15@soton.ac.uk; aw11g15@soton.ac.uk

© Springer Nature Switzerland AG 2021
B. Halak (ed.), *Authentication of Embedded Devices*,
https://doi.org/10.1007/978-3-030-60769-2_5

115

5.1   Introduction

Counterfeited ICs can lead to catastrophic consequences, in particular when they are used in critical infrastructure, military applications or in food and medicine industries. These include significant economic losses (e.g. in the order of billion USD per year in the UK [19]), serious security risks from malfunctioning military weapons and vehicles due to counterfeited parts [12], and potentially loss for human lives (e.g. deaths due to contaminated food, such as 2018 E. coli infection). It is therefore of paramount importance to develop and deploy effective strategies for IC counterfeit mitigation to ensure a trustworthy and secure supply chain. One of main factors magnifying the scale of the counterfeit problem is the trend towards globalisation. The latter is driven by the need to cut costs to gain a competitive advantage and resulted in a remarkable growth of outsourcing levels, which in turn led to a significant increase of supply chains complexity because more firms are involved and the chain must be spread over further tiers [24]. Such an evolution of the supply chain structure has brought about a number of serious challenges linked to the problem of counterfeiting:

- **Visibility [11].** The network of buyer-supplier relationships has become more intricate and participants have little to no visibility and control on upstream stages, which makes it harder to assess the integrity of procured ICs.
- **Traceability [17].** Tracking data is fragmented and spread among involved companies, which makes it very challenging to uniquely identify each procured IC and trace its history back to its origin and, in case of incidents, there is a shortage of data that can be used for forensics investigations.
- **Accountability [10].** In such a scenario afflicted by obscurity and lack of information, fraudulent conduct of companies is noticeably facilitated. There is a lack of means to keep organisations accountable for the portion of processing they handle within the supply chain.

Coping with counterfeiting in these IC supply chains calls for a platform integrated throughout the whole chain to reliably record every transition of products between involved companies. The availability of such a ledger would be an effective means to provide any legitimate actor with precise information on what organisations are operating at upstream stages of the chain (*visibility*) and on the history of each procured IC (*traceability*). Moreover, ensuring recorded transactions are truthful and not tampered with is crucial to enable legally binding liability policies (*accountability*). The implementation of such a platform for counterfeit mitigation requires an infrastructure deployed over the considered supply chain, to enable fine-grained monitoring of ICs sold and bought by involved companies.

---

[1]Multistate Outbreak of E. coli O157:H7 Infections Linked to Romaine Lettuce (Final Update), available online https://www.cdc.gov/ecoli/2018/o157h7-04-18/index.html.

It would be infeasible to identify a single specific authority or enterprise eligible for controlling and operating an infrastructure like this, possibly spanning different countries and diverse regulatory frameworks. Furthermore, such an authority should be trusted globally and have the resources to effectively set up and maintain such a worldwide, complex interconnected network, ensuring at the same time top levels of security, availability and performance.

A decentralised approach is more suitable, where the infrastructure itself is a peer-to-peer network distributed across all the supply chain partners, devoid of any centralised control that may become a single point of failure or a performance bottleneck. An emerging technology that lends itself well to implement a platform like that is the *blockchain*, because of its full decentralisation, high availability and strong guarantees on the immutability of stored data. In brief, a blockchain is a distributed system consisting of a network of peer nodes sharing a ledger of transactions, where each peer keeps a replica of that ledger. The consistency among replicas is ensured by a distributed consensus algorithm run by all the nodes, which also guarantees that transactions cannot be censored or redacted unless an attacker succeeded in controlling a certain percentage of nodes or of computational power. In addition to storing data, blockchain can be used to execute application logic through the *smart contract* technology. A smart contract is an application whose code and execution traces are stored immutably in the blockchain, which provides strong guarantees on execution integrity.

Since such infrastructure has to be run across a predefined set of parties, and considering that part of managed data is not meant to be disclosed publicly, it is reasonable to not rely on existing public permissionless blockchains like Ethereum's. Rather, it is more sensible to build on a *consortium blockchain* where nodes are authenticated, membership is predetermined and data cannot be accessed from the outside.

5.2 Chapter Overview

In this chapter, *we introduce Anti-BlUFf (Anti-counterfeiting Blockchain- and PUF-based infrastructure), an approach based on consortium blockchain and smart contract technologies for item tracking and counterfeit detection in IC supply chains.* Items, i.e. ICs, are uniquely identified to enable tracking by using tamper-proof tags. We choose to use *physically unclonable functions* (PUF) to implement those tags. PUFs are circuits that provide unique signatures deriving from manu-facturing process variations of the circuits themselves. Each alteration of those tags leads to changes of the function computed by the PUF, hence this technology is well suited to enable counterfeit detection. We provide the design of a supply chain management system based on the proposed approach and carry out a preliminary analysis on its effectiveness and feasibility. We define the adversary model to characterise what types of threats can arise in the context of supply chain counterfeit. We then analyse how the proposed design can address those threats to deliver improved counterfeit detection. Finally, to show the technical feasibility of this solution, we describe its prototype implementation and preliminary experimental evaluation, where we measure the effectiveness of using PUFs for counterfeit detection. Finally, we provide an ample discussion on some key pragmatic aspects of integrating the proposed platform with real supply chains.

Although some other blockchain-based IC supply chain management systems have been proposed in literature and industry, a few of them rely on PUFs for item tracking. The main novelty of this work lies in presenting a more complete solution that encompasses (i) the integration of PUF and consortium blockchain, (ii) the detailed description of smart contract implementation and how PUF data is stored in the blockchain and (iii) a security analysis against a threat model.

Our Contribution We rely on blockchain, smart contract and PUF technologies to design a tracking system of ICs for supply chain management, aimed at mitigating the problem of counterfeiting. With respect to the state of the art on this topic, our main research contributions are

- the explicit modelling of the overall system, including IC supply chain, blockchain, smart contracts, PUFs and adversary behaviour, i.e. the *threat model*;
- the detailed design of the proposed tracking system for detecting counterfeits in IC supply chains;
- based on the designated threat model, the identification of the possible attacks to the tracking system aimed to bypass counterfeit detection;
- the analysis of how the proposed tracking system reacts against each of the identified attacks;
- a prototype implementation and preliminary experimental evaluation of the proposed tracking system, where PUF-based counterfeit detection accuracy is assessed;
- a discussion on most relevant points concerning the integration of our solution in real scenarios.

Chapter Organisation The remainder of this chapter is organised as follows. Section 5.3 describes related work. Section 5.4 introduces background information on blockchain and smart contract technologies. The system model is presented in Sect. 5.5, as well as the threat model. Our tracking system is detailed in Sect. 5.6 and its security properties are analysed in Sect. 5.7. Section 5.8 describes the prototype implementation and evaluation. Section 5.9 discusses security analysis results and the limitations of our solution. Finally, Sect. 5.10 outlines conclusion and future work.

5.3 Related Work

The use of blockchain and smart contracts for supply chain management is currently being investigated in some recent industrial projects,[2] and led to the launch of a number of new businesses and companies, which supports the perceived potentialities of this application. Some of these projects use a blockchain-as-a-service solution provided by a third party, such as TradeLends,[4] which employs the platform delivered by IBM Cloud. The limitation of such an approach is the need to totally trust an external organisation, which brings about the same issues mentioned before regarding centralisation.

Different companies use diverse technologies to tag products and reliably link physical assets to the blockchain. Waltonchain[5] uses RFID (Radio-frequency identification) as tags to identify and track items along the chain. Others make use of proprietary solutions. For example, BlockVerify[6] uses their own Block Verify tags, Chronicled[7] employs trusted IoT chips, Skuchain[8] applies Proof of Provenance codes called Popcodes. The problem of existing approaches that rely on the use of RFID-based tags is that these tags are vulnerable to cloning attacks [13, 15], this makes it less effective in protecting against counterfeit attempts.

RFID are also proposed by Toyoda et al. [21]. They introduce a blockchain-based solution for product ownership management system, to be used to prevent counterfeits in the post supply chain. They explain how their system allows to detect counterfeits, and discuss the provided security guarantees only in terms of the possible vulnerabilities of the underlying technology they use, i.e. Ethereum.[9]

Alzahrani and Bulusu [2] propose a solution based on Near Field Communication (NFC). They present Block-Supply Chain, a design for a consortium blockchain-based supply chain where products are tracked using NFC technology to detect counterfeits. Their security analysis is limited to the novel consensus protocol they propose and does not take into account any other aspect of the overall supply chain ecosystem, which includes, but is not restricted to, the blockchain. Furthermore, they do not define a threat model to specify what attacks they want to defend from.

---

[2] How Blockchain Will Transform The Supply Chain And Logistics Industry (https://www.forbes.com/sites/bernardmarr/2018/03/23/how-blockchain-will-transform-the-supply-chain-and-logistics-industry).

[3] Using blockchain to drive supply chain transparency (https://www2.deloitte.com/us/en/pages/operations/articles/blockchain-supply-chain-innovation.html).

[4] TradeLends, available online https://www.tradelens.com/.

[5] Waltonchain https://www.waltonchain.org/doc/Waltonchain-whitepaper_en_20180208.pdf.

[6] BlockVerify: Blockchain-Based Anti-Counterfeit Solution, Introducing transparency to supply chains http://www.blockverify.io/.

[7] Chronicled: Trusted Internet of Things and Smart Supply Chain Solutions, Secure identities, trusted IoT data and automated business logic https://www.chronicled.com/.

[8] Skuchain: Turn Information Into Capital http://www.skuchain.com/.

[9] Ethereum Project (https://www.ethereum.org/).

We propose to produce tamper-proof tags by using physically unclonable functions (PUF), i.e. circuits that can generate a unique identifier for each chip due to the intrinsic variability of the IC fabrication process. Previously reported works on using PUF technology in the context of IC supply chain management are limited in both scope and depth. Guardtime [9] proposes the use of PUF for IoT device authentication based on a consortium blockchain (i.e. KSI Blockchain). However, they provide no clear information on the integration with supply chain they do not explain how PUF data is stored and do not provide any security analysis.

Islam et al. [14] propose the use of PUF and consortium blockchain for tracing ICs. Their work does not investigate in depth what security guarantees are provided and gives no description of the way PUF data is stored in the blockchain.

Similarly, Negka et al. [18] describe a method to detect counterfeit IoT devices by tracking each single device component along the supply chain. They rely on PUFs to authenticate components and implement their detection logic in Ethereum. Although they provide some figures on the fees to pay to use Ethereum smart contracts, they do not detail how PUFs and smart contracts are integrated, nor what specific mechanism is actually employed to implement the detection. Obtained detection accuracy and provided security guarantees are not discussed.

To the best of our knowledge, the lack of appropriate security analysis of the proposed solutions is currently a gap in the state of the art on the application of blockchain and PUF technologies for counterfeiting mitigation in IC supply chains. Table 5.1 details how our solution, Anti-BlUFf, compares with respect to the related work considered in this section. Anti-BlUFf is the only proposed approach that at the same time (i) relies on PUF and consortium blockchain, (ii) gives details on smart contract implementation and how PUF data is stored in the blockchain and (iii) includes a security analysis against a threat model.

5.4   Blockchain and Smart Contract

A blockchain is a decentralised ledger of transactions, fully replicated over a number of trust-less nodes organised in a peer-to-peer network, i.e. the *blockchain network*. Generally speaking, decentralisation is a process by which the operations of a system or organisation, usually those regarding decision-making and planning, are delegated away from a single, trustworthy component or group. In the context of blockchain, decentralisation means that there is no single organisation with a controlling role that can administer the whole system. Transactions represent events of interest for the specific application scenario. For example, in Bitcoin a transaction represents an exchange of a certain amount of cryptocurrency. The name of blockchain derives from the data structure used to store transactions. As shown in Fig. 5.1, transactions are grouped in blocks and added periodically over time to the ledger. Each block includes a number of confirmed transactions and a link to the previous block of transactions. The link is the hash of the content of the previous

Table 5.1 Comparison of Anti-BlUFf with state of the art in blockchain-based anti-counterfeit approaches for supply chains

Proposed solution	Tag type	Blockchain type	Counterfeit detection approach	Security analysis
Toyoda et al. [21]	RFID	Ethereum	Smart contracts pseudo-code provided	Yes
Block-Supply Chain [2]	NFC	Consortium	No details are provided on how the smart contract is implemented	No
Guardtime [9]	PUF	Consortium	No details on integration with blockchain, no info on how PUF data is stored in the blockchain, no details are provided on how the smart contract is	No
Islam et al. [14]	PUF	Consortium	No info on how PUF data is stored in the blockchain, no details are provided on how the smart contract is implemented	No
Negka et al. [18]	PUF	Ethereum	No info on how PUF data is stored in the blockchain, no details are provided on how the smart contract is implemented	No
Anti-BlUFf	PUF	Consortium	Smart contract pseudo-code provided, as well as details on how PUF data is stored in the blockchain	Yes

Fig. 5.1 Graphical representation of the way blockchain transactions are organised in blocks

block. Hence, the ledger is represented as a chain of blocks of transactions, which grows in length over time as new transactions are submitted and confirmed.

Transactions are submitted to the blockchain network and stored in the ledger. A consensus algorithm is run among blockchain nodes to guarantee the consistency of the ledger, in terms of what transactions are included and their order. A blockchain provides strong guarantees in terms of *availability*, because a peer-to-peer network with several nodes and no single-point-of-failure is used. Indeed, all the blockchain nodes should be taken down in order to make the blockchain as a whole unavailable, which may require an adversary to target thousands of different machines. Furthermore, as the ledger is replicated and several nodes participate in the consensus algorithm, an adversary should take control of a relevant fraction of nodes to take over the blockchain and tamper with the ledger. That fraction of nodes depends on the chosen consensus algorithm. This feature lets the blockchain also provide strong guarantees in terms of *integrity*, because the effort an adversary should put to violate the integrity of the ledger is significant.

In open, *permissionless blockchains* like Bitcoin[10] and Ethereum, any node can join and leave the network without any form of authentication, hence additional mechanisms are required to cope with this variability of network membership and the potential presence of malicious nodes. Transactions are submitted to the blockchain network and nodes propagate them so that as many blockchain nodes as possible receive all submitted transactions. Periodically, a new block is created, including a list of confirmed transactions. Any node in the blockchain network can take part in the creation of blocks. The process of creating new blocks is referred to as *mining*, and a node taking part in the mining is referred to as *miner*.

Proof-of-Work (PoW) is commonly employed in this type of permissionless blockchains to regulate how new blocks are mined. Each miner chooses the transactions to include in a block, among those it has received so far that have not been confirmed yet. Furthermore, the miner also chooses the previous block, i.e. the existing block where the new block is meant to be added. Then, the miner works on resolving a resource-intensive puzzle to compute the PoW of that block. Computing the PoW of a block consists in finding a number such that the hash of the whole block has at least a certain number of zeros as most significant digits. The PoW is hard to compute and very simple to verify. This is done to avoid that any node can propose as many new blocks as it would like, overloading the network to the point it becomes unavailable. The difficulty of computing the PoW can be tuned by increasing or decreasing the required number of zeros. In Bitcoin, the difficulty is tuned periodically to ensure that a new block is generated about every 10 minutes. Once the puzzle is solved, the new block is broadcasted to the network and each node can add it to its local blockchain replica. PoW mining almost ensures that one block at a time is generated and spread over the network. Forks are really unlikely to occur, and in case of fork a single chain is eventually restored as miners are expected to always mine a new block over the last block they received.

The probability to generate a new block is proportional to the hashpower (number of hash computed per second) of a miner. A miner controlling the majority of the hashpower would allow that miner to generate blocks of its choice over a branch that will eventually become the longest, hence the network would no longer be decentralised. The immutability property of the blockchain derives from the practical unlikelihood for a single miner to have the majority of the hashpower. Although effective in countering cyber threats stemming from malicious blockchain nodes, PoW is time-consuming and greatly restricts performance [8].

---

[10]Bitcoin (https://bitcoin.org/en/).

In *consortium blockchains* like Hyperledger Fabric,[11] blockchain membership is restricted to the nodes owned by interested organisations, so that each involved firm can take part to the overall process and no external actor can interfere with any operation or read any exchanged data. In this way, blockchain nodes are known and can be reliably authenticated. This also allows to replace PoW with other, more efficient techniques that ensure high-level performance in terms of latency and throughput, such as byzantine fault tolerance algorithms [5].

On top of a blockchain, a smart contract execution environment can be built, to extend the functionalities of the blockchain beyond storing data and allow the execution of any application logic. A smart contract is the code implementing the required application logic and it can be installed in a blockchain likewise a normal transaction, which ensures consequently its integrity. A smart contract defines an interface with methods that can be called externally. Each invocation of a smart contract method is stored as a blockchain transaction, hence the execution trace can be considered as immutable. In general, computations executed through smart contracts are fully transparent and tamper-proof.

5.5 System Model

This section defines the system model representing supply chain (Sect. 5.5.1), PUF-equipped items (Sect. 5.5.2), blockchain and smart contracts (Sects. 5.5.3 and 5.5.4, respectively). Finally, thread model is introduced in Sect. 5.5.5.

5.5.1 Supply Chain Model

An IC supply chain SC includes N parties $\mathcal{P} = \{p_i\}$, i.e. organisations involved in the chain with different roles, and that engage among themselves by supplying and buying items, i.e. ICs. A *supplier* is a party that provides items, while a *buyer* is a party that receives items. Each party can act at the same time as supplier for a number of buyers and as buyer for diverse suppliers. There can be parties that are neither suppliers nor buyers for any other party but operate anyway in the supply chain, such as auditors or regulators. This kind of parties usually needs to access tracking data to assess compliance and solve disputes.

We model SC as a directed acyclic graph $(\mathcal{P}, \mathcal{R})$, where $\mathcal{R}$ is the set of binary supplier-buyer relationships holding within SC. Figure 5.2 shows an instance of the supply chain model. Each element of $\mathcal{R}$ is in the form (p_i, p_j), with $p_i, p_j \in \mathcal{P} \wedge p_i \neq p_j$, and represents a supplier-buyer relationship where p_i is the supplier and p_j the buyer. According to these relationships, parties can be organised

---

[11]Hyperledger Fabric (https://www.hyperledger.org/projects/fabric).

Fig. 5.2 Graphical
representation of an instance
of the supply chain model
with 8 parties $p_0, \ldots, p_7$
spread across 3 stages.
Arrows represent the
supplier-buyer relationships,
e.g. (p_1, p_3) models the fact
that p_1 is a supplier of p_3

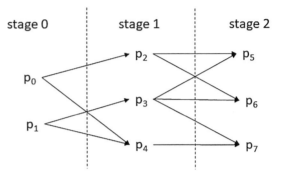

in stages, i.e. the stages of the supply chain. Let S be the number of stages of $\mathcal{SC}$.
Without loss of generality,[12] we define the function $stage \colon \mathcal{P} \to \mathbb{N}$ as follows

$$stage(p) = \begin{cases} 0 & \text{iff } \nexists q \in \mathcal{P} \mid (q, p) \in \mathcal{R} \\ i + 1 & \text{otherwise} \end{cases} \tag{5.1}$$

where $i = \max\limits_{q \in \mathcal{P} \mid (q,p) \in \mathcal{R}} stage(q)$.

Equation 5.1 computes the stage of a party p in the supply chain by
recursively identifying the supplier of p operating at the highest stage, i.e.
$\max\limits_{q \in \mathcal{P} \mid (q,p) \in \mathcal{R}} stage(q)$. Trivially, the stage of p is one unit higher than the stage
of that supplier. If instead p has no supplier (i.e. $\nexists q \in \mathcal{P} \mid (q, p) \in \mathcal{R}$), this
means that p operates at stage 0. Although Eq. 5.1 covers the cases where a buyer
has suppliers in different stages, this is not likely to happen in real supply chains.
Indeed, buyers commonly purchase items from parties in the previous stage only.
Therefore we introduce the following constraint

$$\forall (p, q) \in \mathcal{R} \quad stage(q) - stage(p) = 1 \tag{5.2}$$

We assume the existence of a reliable public key infrastructure (PKI) for the
parties in $\mathcal{P}$. Each party p_i has a key pair (pk_i, sk_i), where pk_i is the public key
known to all the other parties and sk_i is the private key known to p_i only. We discuss
in Sect. 5.9 how such a PKI can be realised and the related issues. Given a key k and
a plaintext message m, we indicate with $|m|_k$ the ciphertext derived from encrypting
m with k. We use $\langle m \rangle_{\sigma_i}$ to indicate that the message m has been signed by p_i, i.e.
that it includes a digest of m encrypted with sk_i.

---

[12]It would be possible for an organisation to operate at different stages of a supply chain. In these
cases, we model such an organisation as multiple parties, one for each stage where it operates.

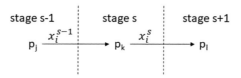

Fig. 5.3 Graphical representation of item x_i moving from p_j in stage $s-1$ to p_k in stage s (x_i^{s-1}), and from there to p_l in stage $s+1$ (x_i^s)

5.5.2   PUF-Equipped Item Model

A number of items are moved along the supply chain $\mathcal{SC}$, from parties at stage 0 to downstream parties. We refer to the generic i^{th} PUF-equipped item produced at stage 0 of the supply chain as x_i, and to the party that produced it as its *producer*. Furthermore, as items can be forged along the chain, we define x_i^s as the item x_i *after its processing* at stage s, where $s = 0 \ldots S - 1$. That is, x_i^s is the item x_i when it is delivered from the supplier at stage s to the buyer at stage $s + 1$ (see Fig. 5.3).

We refer to the function computed by the PUF integrated with item x_i^s as $puf_i^s : \mathbb{N} \to \mathbb{N}$. When an item x_i is produced at stage 0 and equipped with a PUF, it is considered intact.

If x_i is never tampered with along the chain, then the following property holds with high probability

$$\forall c \in \mathbb{N} \quad \forall s \in [1 \ldots S - 1] \quad puf_i^0(c) = puf_i^s(c) \tag{5.3}$$

If instead x_i is forged at stage $s > 0$, then $puf_i^0 \neq puf_i^s$ and the following property holds with high probability[13]

$$\forall c \in \mathbb{N} \quad puf_i^0(c) \neq puf_i^s(c) \tag{5.4}$$

The fact that Eqs. 5.3 and 5.4 do not hold with 100% probability can be accounted for by querying the PUF more times, in order to increase that probability exponentially. We consider the case where PUFs are built by using techniques that mitigate the risk of ML-based attacks, hence we assume that an adversary cannot clone a PUF by collecting a sufficient number of challenge-response pairs.

5.5.3   Blockchain Model

We consider a *consortium blockchain* $\mathcal{B}$ with N nodes $\mathcal{N} = \{n_i\}$, deployed over the supply chain parties' premises (see Sect. 5.4). More precisely, node n_i is located at party p_i. Nodes can communicate among each other over the network

---

[13]Even if the two functions are different, they might return the same response for some challenge.

by sending messages. The network is asynchronous, there is no known bound on message latencies but messages are eventually delivered to their destination. $\mathcal{B}$ uses a *byzantine fault tolerant consensus* protocol, such as PBFT [5], which ensures *safety* if up to $f = \lfloor \frac{N-1}{3} \rfloor$ nodes are byzantine. Section 5.5.5 will explain how byzantine nodes behave.

Interactions between nodes take place by sending digitally signed messages. When a node n_i wants to send a message m to another node n_j, n_i sends a message $\langle i, j, ts, m \rangle_{\sigma_i}$ to n_j. The parameter ts is a timestamp set by n_i, used to avoid replay attacks.

Clients running within supply chain parties' premises can submit transactions to $\mathcal{B}$ by broadcasting them to all $\mathcal{B}$'s nodes. Submitted transactions are eventually confirmed by $\mathcal{B}$ and persistently stored, with strong guarantees on their immutability, i.e. persisted transactions cannot be tampered with or removed unless more than $f = \lfloor \frac{N-1}{3} \rfloor$ nodes are byzantine.

5.5.4 Smart Contract Execution Environment Model

Consortium blockchains like those described in Sect. 5.5.3 can support the execution of smart contracts (see Sect. 5.4), i.e. a smart contract execution environment $\mathcal{SCEE}$ can be built on top of a consortium blockchain $\mathcal{B}$. $\mathcal{SCEE}$ is deployed over the same nodes $\mathcal{N}$ of $\mathcal{B}$.

Smart contracts can be installed in $\mathcal{SCEE}$. A smart contract C includes a number of methods, which can be invoked externally, and a key-value store kvs, which can be accessed internally only, inside those methods. The installation of a smart contract C in $\mathcal{SCEE}$ and every invocation of C's methods are persisted as transactions submitted to the underlying blockchain $\mathcal{B}$. This implies that the application logic encoded by a smart contract cannot be tampered with as long as the underlying blockchain $\mathcal{B}$ guarantees immutability, i.e. unless more than $f = \lfloor \frac{N-1}{3} \rfloor$ nodes are byzantine.

The key-value store of each smart contract provides an interface $set(k, v)$ and $get(k)$ to set and get values for given keys, respectively. Any internal key-value storage kvs relies on the underlying blockchain $\mathcal{B}$ to ensure consistency and immutability of its state. In the specific, each set operation invoked through the $set(k, v)$ method is saved as a transaction in $\mathcal{B}$, hence the whole redo log of the storage is persisted immutably [8]. Furthermore, we assume that a single set operation is allowed for each key, i.e. the value stored for a key cannot be overwritten. In case of overwriting attempt, the set operation returns an error. External applications can also register themselves to receive notifications when specific types of transactions are committed, in order to implement callback-based application logic.

In the considered scenario, there is also the need to verify the identity of the entity that invokes a smart contract method, in order to make sure that the invoker is actually authorised to call the method. We assume that each method invocation includes an additional input parameter that proves the identity of the invoker. In particular, this parameter is the invoker's digital signature of the concatenation of all the other input parameters, plus a timestamp to avoid replay attacks. In the following, we do not explicitly include this additional parameter in the pseudo-code of smart contracts in order to keep them as light as possible. However, we specify what actors are expected to invoke each method, and the corresponding verification is assumed to be carried out by relying on this additional parameter.

5.5.5 Threat Model

The final goal of the adversary is to tamper with items to introduce counterfeit ICs in the supply chain. Hence, it aims at avoiding that counterfeit items are detected to prevent raising suspicion. We assume the existence of a single adversary in the supply chain, Sect. 5.9 encompasses a brief discussion on considering the presence of more independent adversaries.

At *supply chain level* (see Sect. 5.5.1), the adversary can operate at one of the parties, say p_A at stage $stage(p_A)$, with $A \in [0 \ldots N - 1]$. We assume that the adversary cannot control more than one party and cannot alter any supplier-buyer relationship.

At *item level* (see Sect. 5.5.2), the adversary can tamper with items during the manufacturing processes of the party p_A where it operates. For each bought item $x_i^{stage(p_A)-1}$, the adversary can decide whether or not to forge it before supplying it in turn to some other party. However, any tampering with $x_i^{stage(p_A)-1}$ affects the internal structure of the integrated PUF, hence $puf_i^0 \neq puf_i^{s_A}$ (see Eqs. 5.3 and 5.4). Furthermore, if the adversary succeeds to collect at least N_{PUF} challenge-response pairs, it can build a clone of the PUF and attach it to a different item, i.e. it can replace an original product with a counterfeit.

At *blockchain and smart contract execution environment levels* (see Sects. 5.5.3 and 5.5.4), the adversary can control the local node n_A of $\mathcal{B}$ and $\mathcal{SCE}$, i.e. such node is byzantine. The behaviour of a byzantine node can deviate arbitrarily from the expected conduct, hence it can, for example, drop messages and send not expected or wrong messages. Anyway, the adversary cannot break used cryptographic protocols, hence it cannot decrypt messages encrypted without knowing the corresponding keys and cannot forge message signatures.

5.6 Tracking System

Items are tracked as they move along the supply chain, first when they are produced at stage 0 and then each time they are supplied to a buyer operating at the next stage. When delivered at buyer side, the integrity of each item is verified by using its integrated PUF. Tracking information are stored as blockchain transactions to ensure they are immutable and available to any party in $\mathcal{P}$.

The tracking system is built as a smart contract $\mathcal{TS}$ on top of a blockchain-based smart contract execution environment $\mathcal{SCEE}$ (see Sect. 5.5.4). We consider a consortium blockchain $\mathcal{B}$ like the one presented in Sect. 5.5.3, and leverage on the PUFs integrated with the items to assess whether they have been tampered with (see Sect. 5.5.2). The high-level architecture is shown in Fig. 5.4, where basic building blocks and interfaces with supply chain business processes are highlighted. Consortium blockchain $\mathcal{B}$, smart contract execution environment $\mathcal{SCEE}$ and tracking system $\mathcal{TS}$ are distributed and deployed over the IT infrastructures of all the parties.

Module 1 shows the pseudo-code of the tracking system, which defines the five methods shown in Fig. 5.4. These methods are used to integrate the proposed tracking mechanism with the business processes of the supply chain. In particular, this integration occurs on three specific events: when an item is first introduced in the supply chain at stage 0 (*event 1*, see Sect. 5.6.1), when a supplier ships an item to a buyer (*event 2*, see Sect. 5.6.2) and when an item is verified by a buyer (*event 3*, see Sect. 5.6.3). After an item has been processed by a party in the last stage, no further tracking is enforced. However, consumers can still verify items they buy asking the corresponding producers to release additional batches of CRDs.

All tracking data are kept in the blockchain-based key-value storage via set operations, where any relevant information is digitally signed (see Sect. 5.5.1) by the party executing the method where the set operation itself is invoked. This, together with the constraint that keys cannot be overwritten and method invocations are authenticated (see Sect. 5.5.4), ensures that an adversary cannot execute any tracking system method on behalf of another party.

In order to integrate the business processes of supply chain $\mathcal{SC}$ with the tracking system $\mathcal{TS}$, an additional layer is required to interface the existing legacy business

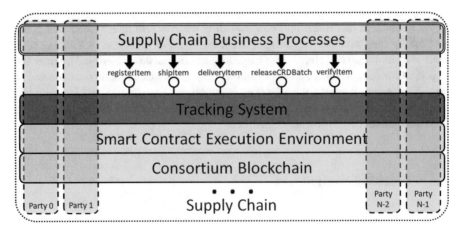

Fig. 5.4 High-level architecture of the tracking system and its positioning within the supply chain

---

Module 1 Tracking mechanism

---

global variables:
1: C ▷ number of challenges to send for each verification
2: R ▷ number of responses (out of C) that need to be correct for the verification to succeed
3: kvs ▷ local key-value storage instance

    ▷ *This method is called by the item producer; p is the item producer identifier, i is the item identifier, $hashedCRD_i$ its hashed challenge-response data*
4: **method** REGISTERITEM($p, i, hashedCRD_i$)
5:     $kvs.set(\langle registered, p, i\rangle, crd_i)$
6: **end method**

    ▷ *This method is called by the seller; p is the item producer identifier, i is the item identifier, s is the seller identifier and b the buyer identifier*
7: **method** SHIPITEM(p, i, s, b)
8:     **if** $kvs.get(\langle registered, p, i\rangle) == null$ **then**
9:         $kvs.set(\langle notRegistered, p, i\rangle, \langle p, i\rangle)$
10:     **else if** $s \neq p \wedge kvs.get(\langle deliver, p, i, s\rangle) == null$ **then**
11:         $kvs.set(\langle notDelivered, p, i, s\rangle, \langle p, i\rangle)$
12:     **else**
13:         $kvs.set(\langle shipped, p, i, s, b\rangle, \langle p, i\rangle)$
14:     **end if**
15: **end method**

    ▷ *This method is called by the buyer; p is the item producer identifier, i is the item identifier, s is the seller identifier and b the buyer identifier*
16: **method** DELIVERYITEM(p, i, s, b)
17:     **if** $kvs.get(\langle registered, p, i\rangle) == null$ **then**
18:         $kvs.set(\langle notRegistered, p, i\rangle, \langle p, i\rangle)$
19:     **else if** $kvs.get(\langle ship, p, i, s, b\rangle) == null$ **then**
20:         $kvs.set(\langle notShipped, p, i, s, b\rangle, \langle p, i\rangle)$
21:     **else**
22:         $kvs.set(\langle delivered, p, b, i\rangle, s)$
23:     **end if**
24: **end method**

    ▷ *This method is called by the item producer; p is the item producer identifier, i is the item identifier, w is the batch index and $crdBatch$ the wth batch of challenge-response pairs for item i, where each response is hashed*
25: **method** RELEASECRDBATCH($p, i, w, crdBatch_{i,w}$)
26:     **if** $w > 0 \wedge kvs.get(\langle crdBatchReleased, p, i, w-1\rangle) == null$ **then**
27:         $kvs.set(\langle invalidBatchID, p, i, w\rangle, crdBatch_{i,w})$
28:     **else**
29:         $hashedCRD_i = kvs.get(\langle register, p, i\rangle)$
30:         **if** $hashedCRD_i[w].hashedC_w \neq hash(crdBatch_{i,w}.challenges)$ **then**
31:             $kvs.set(\langle invalidBatch, p, i, w\rangle, crdBatch_{i,w})$
32:         **else**
33:             $kvs.set(\langle crdBatchReleased, p, i, w\rangle, crdBatch_{i,w})$
34:         **end if**
35:     **end if**
36: **end method**

▷ *This method is called by the buyer; p is the item producer identifier, i is the item identifier,*
s is the seller identifier and b the buyer identifier, w is the batch index and crd Responses$_{i,w}$
is the vector with the C PUF responses
37: **method** VERIFYITEM($p, i, s, b, w, crd Responses_{i,w}$)
38:     **if** $kvs.get(\langle verify, p, i, s, b, w \rangle) \neq null$ **then**
39:         $kvs.set(\langle batchAlreadyVerified, p, i, w \rangle, \langle p, i \rangle)$
40:     **end if**
41:     $crdBatch_{i,w} = kvs.get(\langle crdBatchReleased, p, i, w \rangle)$
42:     **if** $crdBatch_{i,w} == null$ **then**
43:         $kvs.set(\langle noBatchReleased, p, i, w \rangle, \langle p, i \rangle)$
44:     **else**
45:         $correctResponses = 0$
46:         **for** $y = 0$ **to** $C - 1$ **do**
47:             **if** $crdBatch_{i,w}.hashedResponses[y] == hash(crdResponses_{i,w}[y])$ **then**
48:                 $correctResponses + +$
49:             **end if**
50:         **end for**
51:         **if** $correctResponses < R$ **then**
52:             $kvs.set(\langle verify, p, i, s, b, w \rangle, FAIL)$
53:         **else**
54:             $kvs.set(\langle verify, p, i, s, b, w \rangle, SUCCESS)$
55:         **end if**
56:     **end if**
57: **end method**

process management software of SC with the TS smart contract. This integration can be achieved through standard software engineering approaches and does not entail any element of novelty or challenge, so it is not described here. However, this integration layer needs to be accounted for as another potential attack surface that the adversary may exploit, hence in Sect. 5.7 we also address the corresponding security implications (Attack 4).

5.6.1 Event 1: New Item

When a new item x_i is produced by a party p_j at stage 0, a PUF is integrated with x_i and $B \times C$ challenge-response pairs $\langle c_k, r_k \rangle$ are collected. C challenges will be used for each item verification, which makes it more robust against possible variations in the responses generated by a PUF. Hence, up to B parties can verify the integrity of an item at delivery time. B has to be set sufficiently large to accommodate for verifications requested by supply chain parties, end users and external auditors.

The set of pairs is partitioned in B disjoint batches b_w, with $w = 0 \ldots B - 1$, each containing C pairs. Each challenge-response pair for batch w is produced by generating a unique random challenge $c_{w,k} \in \mathbb{N}$, giving it as input to the PUF of x_i and recording the corresponding output $r_{w,k} = puf_i^0(c_{w,k})$. We refer to the vector of batches of challenge-response pairs as the *challenge-response data* CRD_i of x_i, i.e. $CRD_i = [b_0, \ldots, b_{B-1}]$.

CRDs are not disclosed forthwith to all the other parties, otherwise an adversary could develop an ad-hoc circuit to provide correct responses to expected challenges, which could then be used to introduce counterfeits. Rather, at this stage the producer discloses a hashed version of CRD_i, referred to as $hashedCRD_i$, which is a vector of B pairs $\langle hashedC_w, hashedR_w \rangle$, where $hashedC_w = hash(c_{w,0}, \ldots, c_{w,C-1})$ and $hashedR_w = hash(r_{w,0}, \ldots, r_{w,C-1})$, i.e. each pair contains (i) the hash of the concatenation of all the challenges of the batch and (ii) the hash of the concatenation of all the responses of the batch. The method $registerItem()$ is invoked after the generation of the CRD. This method simply stores in the key-value storage the information that $hashedCRD_i$ is available and has been produced by party p_j (line 5). Furthermore, p_j registers itself to be notified (see Sect. 5.5.4) whenever a delivery transaction for x_i is stored into the blockchain (see Sect. 5.6.3).

In order to prevent that any two items in the whole supply chain could clash in the key-value store, the key used to store hashed CRDs also includes the producer party's identifier. The latter has to ensure that no two items are assigned the same identifier among those it registers .

5.6.2 Event 2: Item Shipping

When a party p_s finishes the manufacturing processes of an item $x_i^{stage(p_s)}$ and supplies it to a buyer p_b operating at the next stage, the procedure $shipItem()$ is invoked. Likewise $registerItem()$, this method simply tracks in the blockchain the fact that item x_i, produced by party p_p, has been shipped from party p_s to party p_b. At line 13 of Module 1, all the relevant shipping information are included in the key to make it easier to retrieve shipping data. The value, i.e. the second parameter of the set operation, is not significant and is set to $\langle p, i \rangle$ by convention. Indeed, when querying the blockchain on whether the shipping of item x_i, produced by p_p, from party p_s to party p_b took place, it suffices to check that the value stored for the key $\langle shipped, p, i, s, b \rangle$ is not null.

The method $shipItem()$ also queries the key-value store to perform checks regarding the registration of x_i by p_p and, in case $p_p \neq p_s$, whether x_i has been previously delivered to p_s.

5.6.3 Event 3: Item Delivery and Verification

When an item is delivered to a party p_b from a supplier p_s operating at the previous stage, an integrity verification is carried out. This process includes three steps, each corresponding to a different method: (i) the buyer first notifies that the item has been delivered, then (ii) the item producer releases a batch with C challenge-response pairs, where challenges are in clear and responses are hashed and, finally, (iii) the buyer queries the item PUF with those challenges and publishes obtained responses to enable item verification by any party in the supply chain. The following three subsections describe each step in detail.

5.6.3.1 Item Delivery

The buyer p_b acknowledges the reception of x_i by invoking the method $deliveryItem()$, which stores in the blockchain the fact that x_i, produced by p_p and shipped by p_s, has been delivered to p_b. This method also carries out sanity checks to verify the existence of blockchain records proving that x_i was actually produced by p_p and shipped by p_s to p_b. Party p_b also registers itself to be notified (see Sect. 5.5.4) whenever a new batch release transaction for x_i is committed (see next Sect. 5.6.3.2).

5.6.3.2 Challenge-Response Batch Release

The producer p_p of x_i is notified of the delivery and releases a new batch of challenge-response pairs. Party p_p keeps track of how many batches have been already released for x_i and makes sure to select from CRD_i a batch that has not been disclosed before. Let w be the index in CRD_i of the new batch to release. The challenges need to be published in clear to enable the buyer to feed them to the item PUF. The responses need to be hashed instead, to allow to verify whether obtained responses are valid without disclosing the correct responses in clear.

In the specific, p_p prepares a vector $crdBatch_{i,w}$ with C entries, built as follows. Let b_w be the wth batch of CRD_i, i.e. $b_w = [\langle c_0, r_0 \rangle, \ldots, \langle c_{C-1}, r_{C-1} \rangle]$. The kth entry of $crdBatch_{i,w}$ is the pair $\langle c_k, hash(r_k) \rangle$. The method $releaseCRDBatch()$ is invoked by p_p to store $crdBatch_{i,w}$ in the key-value store (line 33).

The sanity checks performed by this method aim to ensure that $w-1$ batches have been already released (line 26) and that the challenges in this batch are consistent with the $hashedCRD_i$ disclosed at item registration time (line 30). To simplify the notation, we introduce the following two convenient fields of $crdBatch_{i,w}$:

- $crdBatch_{i,w}.challenges$ is the concatenation of all the challenges in the batch, i.e. $c_0, \ldots, c_{C-1}$
- $crdBatch_{i,w}.hashedResponses[k]$ is the kth hashed response of the batch, i.e. $hash(r_k)$

5.6.3.3   Item Verification

Party p_b is informed when the batch $crdBatch_{i,w}$ is released. The PUF of item x_i is then queried with the challenges $crdBatch_{i,w}.challenges$ and responses are collected in a vector $crdResponses_{i,w}$. Finally, p_b calls the procedure $verifyItem()$ to disclose obtained responses to the other parties and let them verify whether these responses are valid.

In the specific, this method first verifies that the same batch has not been already verified, in order to avoid replay attacks where an adversary tries to reuse correct responses learned previously (line 38). Then, the $crdBatch_{i,w}$ data is retrieved and the responses provided by the buyer are checked against the hashed responses included in $crdBatch_{i,w}$. If at least R responses out of C are valid, then the verification is considered as succeeded.

5.7   Security Analysis

In this section we discuss what a malicious party p_A operating at stage $stage(p_A)$ can do and how our proposed tracking mechanism would react. We first define the relevant attacks an adversary may launch in Sect. 5.7.1, based on the threat model introduced in Sect. 5.5.5 and the tracking system proposed in Sect. 5.6. Then, in Sect. 5.7.2 we analyse the response of our tracking system to each of the identified attacks and whether it succeeds in coping with them.

5.7.1   Attacks Definition

According to the threat model introduced in Sect. 5.5.5, the adversary p_A can operate at different levels. As it cannot collude with any other party nor control their resources, attacks at *supply chain level* are not relevant. At *item level*, p_A has several options. The basic one is to just forge an intact item before supplying it to another buyer (Attack 1):

Attack 1 The adversary p_A tampers with an item received from an honest supplier and delivers it to an honest buyer at the next stage.   □

If party p_A works at stage 0, it can tamper with an item before its PUF is fed with the required number of challenges to compute the corresponding CRD. In this way, the CRD stored in the blockchain matches the forged item (Attack 2):

Attack 2 The adversary p_A tampers with an item at stage 0 before its CRD is generated and delivers it to an honest buyer at the next stage.   □

At *blockchain and smart contract execution environment* levels, the adversary can try to compromise the application logic of the smart contract or the data stored in the blockchain by properly instructing the local node n_A, i.e. node n_A becomes byzantine (Attack 3).

Attack 3 The adversary p_A alters the behaviour of the local node n_A, i.e. node n_A becomes byzantine.   □

The layer between supply chain business processes and tracking system is an additional attack surface to consider (see Sect. 5.6). At this level, the adversary can compromise the way smart contract methods are invoked, e.g. by using maliciously modified parameters or by not calling a method at all (Attack 4):

Attack 4 The adversary p_A alters how methods of the tracking system smart contract are called.   □

5.7.2 Attacks Analysis

For each of the five attacks identified in the previous subsection, we provide an analysis of how the proposed tracking system reacts.

Analysis of Attack 1 In this scenario, party p_A tampers with an item $x_i^{stage(p_A)-1}$ received by an honest supplier p_s. Since the supplier is honest, we assume that $x_i^{stage(p_A)-1}$ has not been forged yet. We also assume that party p_p, producer of x_i, is honest; we will cover the case where the producer is malicious in the analysis of Attack 2. The tampered item is supplied to another honest party p_j at stage $stage(p_A) + 1$. As p_p and p_j are honest, they comply with the tracking mechanism described in Sect. 5.6; hence, p_j declares it received x_i by invoking the method $deliveryItem(p, i, A, j)$ and p_p releases a new batch of challenge-response pairs for x_i by calling the method $releaseCRDBatch()$. Afterwards, p_j retrieves this batch and uses the included C challenges in clear to query the PUF $puf_i^{stage(p_A)}$ and collect the corresponding responses, which will be used to invoke the method $verifyItem()$ of the tracking system.

We can assume that p_A stored the correct tracking information regarding the shipping of x_i, otherwise an alert discrediting p_A would be raised (Module 1, line 20). We can also assume that the correct CRD of x_i has been stored in the storage, indeed in this scenario we assume the producer of x_i is honest. With reference to Module 1, this means that the check at line 42 is positive and the C PUF responses in $crdResponses_{i,w}$ can be compared against those in $crdBatch_{i,w}.hashedResponses$. From the properties expressed by Eqs. 5.3 and 5.4, and by the fact that $x_i^{stage(p_A)}$ has been tampered with, it follows that, with high probability, less than R out C responses match, hence an alert is raised (line 52) to notify the detection of a counterfeit item supplied by p_A. The accuracy of this forgery detection mechanism clearly depends on the choice of R. In Sect. 5.8 we show an experimental evaluation where R is tuned to maximise the probability that counterfeits are recognised and minimise the chances that intact items are mistaken for forged.

Note that the challenge-response pairs that will be used for the verification are known by the producer party only, hence an adversary could not discover them in advance and build a model to implement a clone.

Analysis of Attack 2 If p_A operates at stage 0 and tampers with an item x_i^0, then there are two cases. If the counterfeiting occurs after the invocation of method $registerItem()$, then this attack is equivalent to Attack 1 and the forgery is detected by the buyer of x_i at stage 1. Otherwise, if the tampering is made before and the stored CRD $hashedCRD_i$ accurately corresponds to puf_i^0, then this attack cannot be detected by the proposed tracking mechanism.

Analysis of Attack 3 The attacker can make the local blockchain node n_A behave arbitrarily, i.e. n_A becomes a byzantine node, with the aim of compromising data stored in the blockchain or the application logic encoded in the smart contract of the tracking system. By design, according to the model presented in Sect. 5.5.3, in a blockchain with N nodes the adversary should control at least $\lfloor \frac{N-1}{3} \rfloor + 1$ nodes to compromise the consensus, hence if there are at least 4 parties in the supply chain, each with its own local blockchain node, then this attack cannot succeed.

Analysis of Attack 4 The adversary can interact with the methods provided by the tracking system differently from what expected. In the specific, p_A can either invoke a method when it should not, or avoid to call a method at all, or purposely specify wrong values for methods parameters. As explained in Sect. 5.6, an adversary cannot call any method on behalf of another party, hence p_A can only operate on the methods it is expected to invoke.

If p_A operates at stage 0, it can intentionally avoid to store the CRD for item x_i, i.e. it can skip calling $registerItem()$ method. The motivation could be to prevent forgery checks from taking place and indeed such a goal can be partially achieved by the attacker. Anyway, the honest party p_j receiving x_i from p_A easily discovers that the required CRD crd_i is missing (line 17) and raises an alert (at line 18). Although no forgery can be actually detected in this way, this alert marks x_i as a suspicious item and p_A as a disreputable party because it did not store the expected CRD.

If p_A does not call method $shipItem()$ when expected, then the next party receiving the corresponding item x_i detects this anomaly at line 19 and consequently raises an alert at line 20, which again explicitly points at p_A as the party responsible for this misbehaviour.

Avoiding the execution of methods $releaseCRDBatch()$ and $verifyItem()$ would bring no advantage to the adversary, with respect to its goal (see Sect. 5.5.5) of introducing counterfeited products without being detected.

Altering the parameters used for either $registerItem()$ or $shipItem()$ method has the same effect of not calling them at all. Altering the parameters of $releaseCRDBatch()$ or $verifyItem()$ methods would not be beneficial for the adversary to introduce counterfeits.

5.8   Experimental Evaluation

We implemented a prototype of the proposed solution to verify the technical feasibility of the integration of blockchain and PUF, and to assess the reliability of PUF technology to accurately detect counterfeit. We used HyperLedger Fabric[14] to implement the consortium blockchain and the smart contract execution environment (see Sects. 5.5.3 and 5.5.4). We chose this platform because it is one of the most stable and well documented platforms for consortium blockchains. The tracking system $\mathcal{TS}$ defined in Module 1 has been coded as a Fabric chaincode. A 4-bit sequential ring oscillator architecture [23] PUF has been synthesised and implemented on 17 separate Zynq Zybo 7000 FPGA boards [6].

The interface between the tracking system and the PUFs has been implemented as a Java application. The communication with PUF has been done using RXTX-Comm,[15] a library which makes use of Java Native Interface (JNI[16]) to provide a fast and reliable method of communication over serial ports. The communication at PUF side has been encapsulated in a dedicated module which used General Purpose Input Output (GPIO) as Tx and Rx pins for Universal Asynchronous Receiver/Transmitter (UART) serial communication.

We first describe how we tuned the PUF (Sect. 5.8.1), then we describe the use case we tested and what results we obtained (Sect. 5.8.2).

5.8.1   PUF Tuning

The tuning of PUFs consisted in choosing the right value of parameter R, i.e. how many responses out of C need to be correct for the validation to succeed, where C is the number of unique challenges sent to the PUF. We set C to 10.

We first generated the CRD for all the 17 PUFs by collecting a large number of challenge-response pairs for each PUF (more than 21000 pairs). We then randomly selected 3 out of the available 17 PUFs for tuning, while the others were used for the prototype test (Sect. 5.8.2). We refer to those 3 PUFs as the tuning PUFs. Challenges drawn from CRD data of all the PUF have been sent to the tuning PUFs to collect the correspondent responses. The resulting dataset has been used to find a value of R that guarantees that each tuning PUF (i) passes the validation when stimulated with its own CRD and (ii) fails the validation when stimulated with CRD of any of the other 16 PUF.

Each tuning PUF has been stimulated with $C = 10$ unique challenges from each of the 17 PUFs (hence including itself) for 15 times. For each batch of C challenge-

---

[14]Hyperledger Fabric (https://www.hyperledger.org/projects/fabric).

[15]RXTXComm (https://seiscode.iris.washington.edu/projects/rxtxcomm).

[16]JNI (https://docs.oracle.com/javase/8/docs/technotes/guides/jni/).

response pairs, different values of R have been tested, ranging from 5 to 9, and the corresponding validation outcome has been recorded. The metrics of interest for the tuning are

- True Admission Rate (TAR): rate of successful validations when the tuning PUF is validated against its own CRD;
- False Admission Rate (FAR): rate of successful validations when the tuning PUF is validated against the CRD of another PUF;
- True Rejection Rate (TRR): rate of failed validations when the tuning PUF is validated against the CRD of another PUF;
- False Rejection Rate (FRR): rate of failed validations when the tuning PUF is validated against its own CRD;

The ideal situation is when TAR and TRR are 1 while FAR and FRR are 0.

Figure 5.5 shows the values of those metrics for R varying from 5 to 9 (out of 10) for the three tuning PUFs. It can be noted that TAR is always 1 and FRR always 0, which means that the tuning PUFs are successfully validated all the times their own CRD is used. When the validation is based instead on CRD of a different PUF, sometimes tuning PUFs still pass the validation. This happens because the functions computed by different PUFs can overlap for certain challenges. Figure 5.5 shows

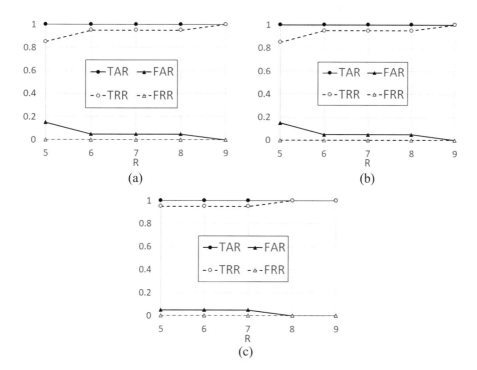

Fig. 5.5 Tuning of 3 PUFs, R is varied between 5 and 9 out of 10 and the corresponding value of metrics TAR, FAR, TRR, FRR are shown

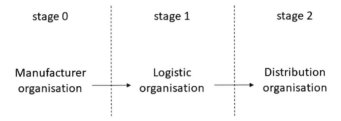

Fig. 5.6 Supplier-buyer relationships in the prototype

that the probability that this occurs (i.e. FAR) decreases as R grows, and that with $R = 9$ FAR is 0 (and TRR is 1) for all the 3 tuning PUFs. Hence, for the prototype test, the validation of a PUF is considered successful if at least 9 out of 10 responses match those stored in the corresponding CRD.

5.8.2 Prototype Test

We developed a prototype with three organisations: manufacturer, logistics and distribution. The corresponding supplier-buyer relationships are depicted in Fig. 5.6. We considered two cases: when no adversary is present and when the logistic organisation is malicious and tampers with the items supplied by the manufacturer before delivering them to the distribution organisation.

We used the other 14 PUFs for the prototype test, 8 for the case where no party is malicious and 6 for the case where the logistic organisation is the adversary. In the latter case, the manufacturer delivers 3 PUFs to the logistic organisation, which replaces each of them using the other 3 PUFs and delivers them to the distribution organisation.

When there is no adversary, all the 8 PUFs pass the validation both at the logistic and at the distribution organisation, hence the TAR is 1 and FRR is 0. When instead the logistic organisation replaces the three PUFs, all of them fail the validation at the distribution organisation, therefore the FAR is 0 and TRR is 1.

These preliminary results are promising to prove both the technical feasibility and the effectiveness in counterfeit mitigation of the proposed tracking system.

5.9 Discussion

This section discusses several key aspects of the proposed solution, pointing out key limitations and main research directions to investigate as future work: the results of the security analysis (Sect. 5.9.1), the issues of implementing a PKI infrastructure for a consortium blockchain (Sect. 5.9.2), the limitations of the chosen threat model (Sect. 5.9.3), the feasibility of embedding PUFs within the items to track (Sect. 5.9.4), possible privacy issues when sharing data among parties through the blockchain (Sect. 5.9.5), observations on consortium blockchain performance and scalability (see Sect. 5.9.6) and, finally, considerations on the costs associated with adopting the proposed solution in real supply chains (Sect. 5.9.7).

5.9.1 Security Analysis Results

The results of security analysis presented in Sect. 5.7 show the capability of the proposed tracking system to be effective against the identified attacks. Any attempt to counterfeit items (Attack 1) is correctly detected and attributed to the right malicious party.

If the adversary operates at stage 0 and tampers with the item before the corresponding CRD is built and stored in the blockchain (Attack 2), then the tracking system fails to detect the forgery. This derives trivially from relying on the CRD itself to be the trust root of the whole counterfeit detection mechanism. Enhancing the proposed approach to cover threats happening before CRD generation is one of our main future work.

The other attacks at software level, to make a blockchain node byzantine (Attack 3), or at the interface between supply chain business processes and tracking system (Attack 4), have been shown to be not effective. On the one hand, this derives from by-design security properties provided by blockchain-based systems, indeed using PBFT-like consensus algorithms allows to tolerate a single byzantine node when the blockchain includes at least four nodes (Attack 3). On the other hand, the tracking system prevents an adversary from invoking smart contract methods on behalf of a different party, so attacks based on altering how methods are called (Attack 4) are not relevant.

5.9.2 PKI Infrastructure for Consortium Blockchains

The proposed tracking system relies on a consortium blockchain (see Sect. 5.5.3), which in turn requires a reliable PKI to obtain the relationships between parties' identities and public keys. These certificates are issued when the platform is set up at the beginning and when the supply chain membership changes. From a security perspective, the PKI is a single-point-of-failure, i.e. an adversary may target the PKI to take over the whole blockchain, and thus the tracking system.

This problem has been already addressed in literature. For example, there exist proposed solutions based on blockchain to decentralise the PKI so as to make it much more resistant to cyber attacks [1, 7], and provide attack tolerance guarantees comparable to those already provided by the tracking system. These solutions are based on public blockchains, which may introduce privacy issues. Other approaches have been proposed for privacy-preserving blockchain-based PKI, such as PB-PKI [3]. The integration of the tracking system with this type of PKI is out of the scope of this work and is left as future work.

5.9.3 Threat Model Limitations

The list of attacks identified in Sect. 5.7.1 depends tightly on the threat model introduced in Sect. 5.5.5, which in turn derives from three main assumptions: (i) there is a single adversary, (ii) it controls exactly one party and (iii) only aims at introducing counterfeits in the supply chain. It can be reasonable to consider the implications of relaxing those assumptions and identify what additional attack scenarios may arise when an adversary can control more parties, when more adversaries are active, either independently or by colluding among themselves, and when the adversary has a different goal.

We can expect that a security analysis of the proposed tracking system against such a stronger attack model would point out further vulnerabilities. For example, an adversary could aim at blaming another party by tampering with an item just after the delivery and before it gets verified by the tracking system. To avoid any attribution, the adversary can blame the corresponding supplier for the shipping of a counterfeit item. However, this analysis should be integrated with a risk assessment to measure the likelihood of more advanced attacks, and should estimate out to what extent they can be considered reasonable. Taking into account wider threat models is an additional potential future work.

5.9.4 Embedding PUFs Within Items to Track

The effectiveness of tracking items by using PUFs strictly depends on how easily an adversary can forge items without affecting the PUFs themselves. If a PUF can be removed from an item and embedded within a different one, then the whole counterfeit detection mechanism is flawed. In the end, this boils down to preliminarily check whether it is technically feasible to embed PUFs within items in such a way that all the properties of the PUF-equipped item model hold true (see Sect. 5.5.2).

Electronic components are items where PUFs can be easily and cheaply implanted by integrating PUF circuitry inside the component circuitry, ensuring that PUFs cannot be removed and replaced. Hence, the approach we propose fits well with integrated circuits and IoT devices supply chains. However, an aspect to be taken into account is that a failure of the PUF circuit is likely to lead to inaccuracies in the counterfeit detection process. Although this problem is intrinsic of any tag-based tracking mechanism, it would be interesting to explore the feasibility and challenges of devising methodologies to distinguish between a counterfeited PUF and a damaged PUF.

5.9.5 Privacy Issues

Although the network of companies involved in the supply chain should be made as transparent as possible to enhance visibility, organisations can be legitimately reluctant to disclose their own supplier network and procurement history to other, possibly competitor firms. What information should be shared needs to be adjusted according to this kind of confidentiality requirements, on a case by case basis. An important applied research direction to investigate, for each target supply chain market, concerns this trade-off between privacy and scope, with the aim to find the sweet spot where information on supplier network and procurement history can be shared smoothly.

A general approach to address those privacy issues is to make each transaction only visible to a specific subset of parties. In the specific, only those parties having some stakes on the item referenced in the transaction should be able to read it, so that visibility can be preserved and limited to interested actors only. With reference to our prototype implementation based on Hyperledger Fabric, we could implement this general approach by leveraging on the concept of *channels* to establish between subsets of nodes. A transaction can be associated with a specific channel to ensure only the nodes in that channel can see its content. Our prototype can be enhanced with privacy-preserving techniques by relying on Fabric channels.

5.9.6 Performance and Scalability

While public permissionless blockchains like Ethereum's are known to provide limited performances in terms of transaction latency and throughput, consortium blockchain can commit thousands of transactions per seconds with subsecond latency [4], also in WAN settings [20]. In terms of scalability, BFT-tolerant algorithms have been proposed in literature that can scale to tens of nodes with minor performance penalties [16, 22], which matches realistic supply chain setting including tens of different organisations.

5.9.7 Platform Integration Costs

Each supply chain works according to specific business processes which may differ significantly from market to market. On the one hand, pinpointing the right abstraction level for the interface provided by the tracking system is crucial to increase the cases where it can be integrated. On the other hand, the integration with those business processes deserves a deeper analysis in terms of security, to figure out whether additional cyber threats can be identified at those integration points (see Attack 4 in Sect. 5.7.1), and cost-effectiveness, to quantify whether and to what extent the benefits of counterfeiting mitigation outweigh the costs to accomplish such a large-scale integration.

In terms of cost-effectiveness, it is to note that relying on consortium blockchains rather public permissionless blockchains allows to cut any cost due to the fees to pay when submitting transactions. Indeed, while supply chain tracking solutions based on Ethereum have a per-transaction cost (e.g. see Negka et al. [18]), submitting transactions in Hyperledger Fabric is totally free.

5.10   Conclusion

In this chapter we presented the design of a tracking system aimed to mitigate counterfeits in IC supply chains. The solution we propose is based on blockchain and smart contract technologies to provide high availability and strong tolerance against integrity attacks to stored data and application logic. We rely on physically unclonable functions to uniquely identify and accurately track ICs along the supply chain. We validate our solution against a specific threat model and find out that it is effective to counter the identified attacks, but an adversary operating at the first stage of the supply chain can bypass the anti-counterfeit mechanism. Finally, we implemented and tested a prototype of the proposed tracking system to prove it is technically feasible and accurate in correctly validating both intact and forged items.

In addition to investigate possible solutions to the limitations discovered in the security analysis, other future works include the integration of a reliable PKI infrastructure within the tracking system and the implications of considering a stronger threat model.

References

1. M. Al-Bassam, Scpki: A smart contract-based PKI and identity system, in *Proceedings of the ACM Workshop on Blockchain, Cryptocurrencies and Contracts, BCC '17*, pp. 35–40 (ACM, New York, NY, USA, 2017). http://doi.org/10.1145/3055518.3055530
2. N. Alzahrani, N. Bulusu, Block-supply chain: A new anti-counterfeiting supply chain using NFC and blockchain, in *Proceedings of the 1st Workshop on Cryptocurrencies and Blockchains for Distributed Systems, CryBlock'18*, pp. 30–35 (ACM, New York, NY, USA, 2018). URL http://doi.org/10.1145/3211933.3211939
3. L. Axon, M. Goldsmith, Pb-pki: A privacy-aware blockchain-based pki, in *Proceedings of the 14th International Joint Conference on e-Business and Telecommunications*. SCITEPRESS (2016)

4. A. Bessani, J. Sousa, E.E. Alchieri, State machine replication for the masses with BFT-smart, in *44th Annual IEEE/IFIP International Conference on Dependable Systems and Networks* (2014)
5. M. Castro, B. Liskov, Practical byzantine fault tolerance, in *Proceedings of the Third Symposium on Operating Systems Design and Implementation, OSDI '99*, pp. 173–186 (USENIX Association, Berkeley, CA, USA, 1999). http://dl.acm.org/citation.cfm?id=296806.296824
6. L.H. Crockett, R.A. Elliot, M.A. Enderwitz, R.W. Stewart, *The Zynq Book: Embedded Processing with the Arm Cortex-A9 on the Xilinx Zynq-7000 All Programmable Soc* (Strathclyde Academic Media, 2014)
7. C. Fromknecht, D. Velicanu, S. Yakoubov, A decentralized public key infrastructure with identity retention. IACR Cryptol. ePrint Archive **2014**, 803 (2014)
8. E. Gaetani, L. Aniello, R. Baldoni, F. Lombardi, A. Margheri, V. Sassone, Blockchain-based database to ensure data integrity in cloud computing environments, in *Proceedings of the First Italian Conference on Cybersecurity (ITASEC17)*, Venice, Italy, January 17–20, pp. 146–155 (2017). http://ceur-ws.org/Vol-1816/paper-15.pdf
9. Guardtime, Internet of Things Authentication: A Blockchain solution using SRAM Physical Unclonable Functions (2017). Available online: https://www.intrinsic-id.com/wp-content/uploads/2017/05/gt_KSI-PUF-web-1611.pdf
10. J. Hartmann, S. Moeller, Chain liability in multitier supply chains? responsibility attributions for unsustainable supplier behavior. J. Oper. Manag. **32**(5), 281–294 (2014). https://doi.org/10.1016/j.jom.2014.01.005. http://www.sciencedirect.com/science/article/pii/S0272696314000060
11. N.O. Hohenstein, E. Feisel, E. Hartmann, L. Giunipero, Research on the phenomenon of supply chain resilience: a systematic review and paths for further investigation. Int. J. Phys. Distrib. Logist. Manag. **45**(1/2), 90–117 (2015)
12. B.T. Horvath, Not all parts are created equal: The impact of counterfeit parts in the air force supply chain. Tech. rep., Air War College, Air University Maxwell AFB United States (2017)
13. J. Huang, X. Li, C. Xing, W. Wang, K. Hua, S. Guo, Dtd: A novel double-track approach to clone detection for RFID-enabled supply chains. IEEE Trans. Emerg. Top. Comput. **5**(1), 134–140 (2017). https://doi.org/10.1109/TETC.2015.2389532
14. M.N. Islam, V.C. Patii, S. Kundu, On IC traceability via blockchain, in *2018 International Symposium on VLSI Design, Automation and Test (VLSI-DAT)* (IEEE, 2018), pp. 1–4
15. R. Jain, D.K. Chaudhary, S. Kumar, Analysis of vulnerabilities in radio frequency identification (RFID) systems, in *2018 8th International Conference on Cloud Computing, Data Science & Engineering (Confluence)* (IEEE, 2018), pp. 453–457
16. F.P. Junqueira, B.C. Reed, M. Serafini, Zab: High-performance broadcast for primary-backup systems, in *2011 IEEE/IFIP 41st International Conference on Dependable Systems & Networks (DSN)* (IEEE, 2011), pp. 245–256
17. Z. Khojasteh-Ghamari, T. Irohara, Supply chain risk management: A comprehensive review. *Supply Chain Risk Management* (Springer, 2018), pp. 3–22
18. L. Negka, G. Gketsios, N.A. Anagnostopoulos, G. Spathoulas, A. Kakarountas, S. Katzenbeisser, Employing blockchain and physical unclonable functions for counterfeit IoT devices detection, in *Proceedings of the International Conference on Omni-Layer Intelligent Systems* (ACM, 2019), pp. 172–178
19. OECD, *Trade in Counterfeit Products and the UK Economy* (OECD Publishing, Paris, 2017). https://doi.org/10.1787/9789264279063-en. https://www.oecd-ilibrary.org/content/publication/9789264279063-en
20. J. Sousa, A. Bessani, Separating the wheat from the chaff: An empirical design for geo-replicated state machines, in *2015 IEEE 34th Symposium on Reliable Distributed Systems (SRDS)* (IEEE, 2015), pp. 146–155

21. K. Toyoda, P.T. Mathiopoulos, I. Sasase, T. Ohtsuki, A novel blockchain-based product ownership management system (poms) for anti-counterfeits in the post supply chain. IEEE Access **5**, 17465–17477 (2017). https://doi.org/10.1109/ACCESS.2017.2720760
22. M. Vukolić, The quest for scalable blockchain fabric: Proof-of-work vs. BFT replication, in *International Workshop on Open Problems in Network Security* (Springer, 2015), pp. 112–125
23. C. Wachsmann, A.R. Sadeghi, Physically unclonable functions (PUFs): Applications, models, and future directions. Synth. Lect. Inf. Secur. Priv. Trust **5**(3), 1–91 (2014)
24. F. Wiengarten, P. Humphreys, C. Gimenez, R. McIvor, Risk, risk management practices, and the success of supply chain integration. Int. J. Prod. Econ. **171**, 361–370 (2016)

Chapter 6
Hardware-Based Authentication Applications

Md Tanvir Arafin and Gang Qu

Abstract Authentication is one of the most fundamental problems in computer security. Implementation of any authentication and authorization protocol requires the solution of several sub-problems, such as secret sharing, key generation, key storage, and secret verification. With the widespread employment of the Internet of Things (IoT), authentication becomes a central concern in the security of resource constraint internet-connected systems. Interconnected elements of IoT devices typically contain sensors, actuators, relays, and processing and control equipment that are designed with a limited budget on power, cost, and area. As a result, incorporating security protocols in these IoT components can be rather challenging. To address this issue, in this chapter, we discuss hardware-oriented security applications for the authentication of users, devices, and data. These applications illustrate the use of physical properties of computing hardware such as main memory, computing units, and clocks for authentication applications in low power on the IoT devices and systems.

Keywords Authentication · Secret sharing · Key generation · Key storage · Secret verification · IoT · Hardware-based security

6.1 Introduction

Ubiquitous deployment of sensors, actuators, data acquisition systems, processing modules, cloud servers, and electronic components interconnected by wired and wireless communication technology is leading to the era of the Internet of Things (IoT). Implementation of Internet of Things is critically dependent on a well-

Md. T. Arafin
Morgan State University, Baltimore, MD, USA
e-mail: mdtanvir.arafin@morgan.edu

G. Qu (✉)
University of Maryland, College Park, MD, USA
e-mail: gangqu@umd.edu

© Springer Nature Switzerland AG 2021   145
B. Halak (ed.), *Authentication of Embedded Devices*,
https://doi.org/10.1007/978-3-030-60769-2_6

designed interconnected network between the Things. As the number of Things grows, constraints on the critical factors such as cost, area, and power, *etc.* become tighter, and in most cases forces the designer to either sacrifice some design aspects or innovate new design to meet all the requirements. Unfortunately, one of the standard sacrifices is made on the security of the low power components in the network such as the sensors, data acquisition units, or on-field data processing units [1].

Compromising regarding security for low power modules can pose severe threats to the complete system. Therefore, the design should opt for Internet-of-Trusted-Things, where the system components are trusted and provide accurate data to the system. Authentication creates the first layer of trust between two entities—the provers and the verifier. Conventional authentication methods using low-entropy user-passwords are weak and severely vulnerable to dictionary attacks. Although password-authenticated key exchange protocols can render secure authentication against active and passive attackers, they are susceptible to dictionary attacks at the authentication server. On the other hand, authentication secrets can be created and distributed using secret sharing schemes to resist dictionary attacks at the server-side. However, general secret sharing mechanisms such as the Shamir's secret sharing algorithm and its derivatives are hardware independent, and require significant power and computation capabilities. Thus, secure classical authentication strategies can become unfeasible for low power modules where the claimants can constitute low power sensor nodes that do not have enough processing capability to realize essential cryptographic functions.

Hardware security is an emerging field in computer security that studies the application of security primitives in hardware and their vulnerabilities. As new technologies emerge, the nature of security primitives and vulnerabilities change, hence, studying security features of such technologies has its unique benefits. Moreover, new technologies can provide exclusive hardware features useful for security purpose. Instead of making hardware the weakest link in security (for example, side channel attacks), one can employ hardware's intrinsic properties to enhance security. As an example of hardware-based emerging security applications, in this chapter, we discuss several lightweight hardware intrinsic security solutions to solve problems associated with authentication of entities.

6.2 Organization

In this chapter, we will discuss several examples of hardware-based application that authenticates devices, users, and broadcast signals used for navigation. We start with the backgrounds on the fundamentals of hardware-based authentication in Sect. 6.3. In the three subsequent sections we explore three different hardware-based authentication applications. Section 6.4 explores the application of approximate computation in authentication technique and presents a device authentication protocol VOLtA. Section 6.5 gives a detailed overview of single and multi-user authentication using novel memory component: the memristor. Finally, Sect. 6.6 demonstrates the application of crystal oscillators in hardware-based signal authentication applications.

6.2.1   Notations

In this chapter, the set of integers (modulo an integer $q \geq 1$) is denoted by $\mathbb{Z}_q$. Matrices, vectors, and single elements over $\mathbb{Z}_q$ are represented by consecutively upper case bold letter, lower case bold letters and lower case letters such as $\mathbf{X}$, $\mathbf{x}$ and x. For a vector $\mathbf{x}$, the length of the vector is denoted by $|\mathbf{x}|$, ith element is represented by $\mathbf{x}[i]$, and $wt(\mathbf{x})$ denotes the Hamming weight (i.e., the number of indices for which $\mathbf{x}[i] \neq 0$) of the vector $\mathbf{x}$. The Hamming distance between two binary matrices is denoted by $hd(\mathbf{A}, \mathbf{B})$ (i.e., the number of indices for which $\mathbf{A}[i][j] \neq \mathbf{B}[i][j]$). Concatenation of two vectors is represented by $||$ symbol. $c \xleftarrow{\$} \{\mathbf{x} \in \mathbb{Z}\}$ represents a random sampling of $\mathbf{x}$. We denote *probabilistic polynomial time* (PPT) algorithms with upper case calligraphic alphabets such as $\mathcal{A}$. Therefore, if $\mathcal{A}$ is probabilistic, then for any input $\mathbf{x} \in \{0, 1\}^*$ there exists a polynomial $p(.)$ such that the computation of $\mathcal{A}$ terminates in at most $p(|\mathbf{x}|)$ steps.

6.3   Background

6.3.1   Fundamentals on Authentication

Identity and entity authentication depend on the sharing a secret between two parties—the verifier and the prover. In this section, we introduce common secret sharing and user authentication mechanisms.

Single Entity Authentication   For authenticating a single user/entity, the verifier authenticates the singular prover based on a secret that can be derived from (1) something that is known by both parties (such as passwords), or (2) something possessed by the prover (such as hardware keys), or (3) something inherent (such as signatures, biometric signals, etc. of the prover) [2].

Passwords are the most commonly used entity authentication mechanism where the authenticator stores the (username, password (or its hashed value)) pair for different prover and use this pair to identify a prover. To strengthen this protocol several steps can be taken such as (i) different password rules can be introduced, (ii) password salting can be performed, or (iii) password mapping can be slowed down which will make it difficult for an attacker to test a large number of trial passes [2]. Frequent attacks on password schemes are replay attack and exhaustive and dictionary password search. Furthermore, leaking of an authenticator's database containing (username, password)-pairs can cause significant threat [3].

For all of these weaknesses of password schemes, challenge-response identification becomes a step toward strong authentication where the authentic parties share a sequence of secret one time passwords or challenge-response pairs which are usually derived from some one-way functions or a challenge-response (CRP) table. Furthermore, a zero-knowledge protocol that verifies an entity through its possession of the knowledge of the secret instead of the exact secret is also a strong authentication scheme [2].

Secret Sharing and Authentication of Multiple Entities   Secret sharing is a well-studied problem which seeks to distribute pieces of information (or called the secret) to multiple parties in a way such that the information (the secret) can be revealed when all or a sufficiently large subset of the individuals contribute their shares. Shamir's secret sharing algorithm [4] defines the concept of threshold scheme for secret sharing. This (k, n) threshold scheme (where $k \leq n$) can be described as follows:

> **Definition 6.1**   Assume that a secret S to be shared by n parties. The secret is divided into n pieces such that, the knowledge of k or more pieces would be sufficient to reconstruct S. However, if the knowledge of any $k - 1$ pieces or less is available, it would be impossible to restore S.

A direct application of this problem is the multi-entity authentication problem, where at least k authentic users must present their shares to gain access to a system. One may argue that authenticating each user and counting the authenticated users can trivially solve the multi-entity authentication problem. This approach does address the issue but has two significant drawbacks in scalability and user privacy. First, the expensive authentication protocol has to be applied at least k times, and some mechanism to check duplicate users must be implemented. Second, unlike the traditional secret share dependent approaches [4–7], this method will reveal the identity of each authenticated user, creating user privacy concerns.

Shamir [4] and Blakley [5] independently proposed solutions for the aforementioned problem of (k, n)-threshold scheme. More specifically, Blakley used techniques from finite geometry to provide a solution for safeguarding and sharing cryptographic keys. Shamir's solution is designed on the polynomial interpolation over a finite field. It has since become a widely acceptable solution for secure secret sharing and distribution of cryptographic keys. A detailed explanation of the scheme and the impact of subsequent works can be found in [6]. One of the drawbacks of these early solutions is their high computational cost. For example, Shamir's secret sharing algorithm requires calculations over the finite field during both secret share generation phase and secret reconstruction phase. This, in turn, requires complex digital circuitry for hardware implementation, which is known to be more efficient than the software implementation.

6.3.2 Hardware in Authentication

The first problem for authentication is key generation. Assume a verifier Alice tries to authenticate/identify a prover Bob. For solving this authentication problem, a signature of Bob is required to be stored by Alice to verify Bob. This procedure poses several security challenges. First, as discussed in the previous section human-generated keys such as passwords are rarely random and are vulnerable to dictionary attacks. One can use machine-generated private keys. However, machine-generated keys are difficult to remember, and storing such keys in non-volatile memory can easily leak this secret key to an adversary. Hardware-based security primitives such as physically uncloneable functions (PUFs) provide some solution to these problems.

Entity Authentication Using Physically Uncloneable Functions Silicon Physically Uncloneable Functions (PUFs) are on-chip circuitry that can extract fabrication variations to generate chip-dependent PUF data that can be used as secret keys or as seeds of random number generators [8–10]. For using PUFs in authenticating a device, one can hash a password with the physically uncloneable device signature to increase the entropy of the private keys and generate a unique device dependent key. Moreover, PUFs do not store the key; rather the key exists when the device is powered on, and therefore, to extract(or leak) a key an adversary needs to attack the system while it is up and running, which is significantly harder [11]. Thus, hardware-based random number generators and PUFs can be useful in solving primary key generation and storage challenges. Moreover, physical randomness in these PUF designs is considered inherently independent and identically distributed. Therefore, learning attacks that focus on modeling the internal operation of a well-designed PUF fails. The key generation and authentication model for general PUFs are given below.

Algorithm 6.1 PUF based key generation

1: **procedure** $(\mathbf{C}, \mathbf{R}) \leftarrow KeyGen(1^\lambda)$
2:     Assume a PUF can provide ℓ-pairs of challenge response.
3:     Read the PUF and record n-bit responses $\mathbf{R}^{\ell \times n}$ for the p-bit challenges $\mathbf{C}^{\ell \times p}$ to the PUF.
4:     **return** $(\mathbf{C}, \mathbf{R})$

Table 6.1 Single round interactive authentication for protocol using a PUF with high entropy content

Prover(P)	Verifier($\mathbf{C}, \mathbf{R}$)
	Select a random challenge $\mathbf{c} \xleftarrow{\$} \mathbf{C}$
$\xleftarrow{\mathbf{c}}$	
Apply $\mathbf{c}$ to the PUF P and read the response $\mathbf{r}' = PUF(\mathbf{c})$	
$\xrightarrow{\mathbf{r}'}$	
	If $\mathbf{r}' = \mathbf{r}$ (where $\mathbf{r}$ is the recorded response for $\mathbf{c}$), accept, else reject

Enrollment The verifier uses the key generation algorithm to generate the secret $(\mathbf{C}, \mathbf{R})$ from a PUF P. The verifier keeps $(\mathbf{C}, \mathbf{R})$ and the prover owns the PUF.

Authentication An interactive PUF based authentication protocol is given in Table 6.1.

One of the key weakness of such protocol is that the changes in operating conditions and environmental factor can change the physical responses of a PUF system. Therefore, such authentication has high completeness for the systems that have stable PUF bits over all the operating conditions. For example, SHIC PUFs keep the PUF bits as a non-volatile memory content that has an excellent memory retention capability. However, these PUFs are vulnerable to read-out attack and therefore, slowing down of reading access is required for a security guarantee.

Hardware-Based Classical Secret Sharing

On the other hand, a simple hardware-based secret sharing model was first proposed by Naor and Shamir [7] as the Visual Cryptography (VC) scheme. It is an alternative lightweight system for secret sharing to Shamir's original scheme. In visual cryptography, the secret shared among multiple parties is merely an image. The secret image is broken into n pieces, and each piece is printed on a transparency. When k or more pieces of these transparencies are placed on a stack, the secret image is revealed and can be comprehended by human eyes. In this scheme, the secret share generation requires only straightforward calculations, and the revelation of the secret image does not involve any mathematical computation. In essence, this is an example of how inherent physical properties of hardware can be used in designing lightweight security primitives to avoid complex mathematical computation and formal cryptography protocols. Naor and Pinkas first discussed the application of visual cryptography for authentication and identification [12].

6.4 Hardware-Based Authentication Using Approximation Errors

Physical variations in hardware have already been widely utilized for security applications in the form of physically uncloneable function (PUF). However, for device authentication, PUF is costly regarding hardware and power consumption, bringing an obstacle for its widespread usage in IoT. Therefore, in this section, we will illustrate a device authentication protocol that does not require any additional hardware in the low power Things.

The design of digital circuits and systems is focused on the deterministic results, i.e., for the same inputs, any design will yield the same output. As we push the boundaries of power efficiency, we are introducing the effects of analog nature of the circuit components involved in the computation. In the field of approximate computing, one of the common power reduction techniques is voltage over-scaling (VOS). In VOS, the digital circuit used for computation is operated under the nominal voltage, which guarantees correct output for all input conditions under any given operating environment. Since the dynamic power consumed in a VLSI chip is squarely proportional, and static power is proportional to the operating voltage, reducing the operating voltage under the prescribed margin can result in considerable power savings. However, the effect of this voltage over-scaling will be translated into errors generated during a computation. Let us examine how to use voltage over-scaling to exacerbate the effects of process variation and extract information regarding this variation that can be used for security purposes.

Variations in the manufacturing process, supply voltage, and temperature (PVT) have a major impact on the performance and reliability of a computation performed by an integrated circuit. Fluctuations in PVT affect the time required for a gate to switch to the correct state and thus creates timing errors in the output result. The general trend of digital design is to consider all the corner cases and optimize the design in a way so that fluctuations in PVT have no (or minimal) effect on the output under normal operating conditions. As the size of the transistor reduces, the effect of process variation becomes a critical issue in digital design. These variations come from the collective factors such as imperfection of the manufacturing process, random dopant fluctuation, and variation in the gate oxide thickness, etc. As the transistor size shrinks, the standard deviation of threshold voltage variation increases, since it is proportional to the square root of the device area as given by [13]

$$\sigma_{\Delta V_t} = A_{\Delta V_t}/\sqrt{WL} \qquad (6.1)$$

where $A_{\Delta V_t}$ is characterizing matching parameter for any given process. This variation in V_t will have a direct consequence in the delay of a CMOS gate which can be approximated using the following equation [13]:

$$d_{gate} \propto \frac{V_{DD}}{\beta(V_{DD} - V_t)^\alpha} \tag{6.2}$$

where α and β are fitting parameters for a gate and the given process. Therefore, to maintain the timing correctness, static timing analysis of a given design is performed on the process corners. The timing analysis ensures that for a given operating condition, all paths meet the timing requirements to produce correct results irrespective of the input vectors. Scaling V_{DD} from the predefined operating voltage creates large timing errors and degrades the output quality. However, V_{DD} scaling can offer great saving in energy budget, and therefore, voltage over-scaling-based approximate computation received significant research attention in the last decade [14–16].

From Eq. 6.2 it is evident that with voltage over-scaling and transistor shrinking, underlying device signature due to process variation manifests itself more prominently in the delay output. If proper correction mechanism is not applied, this variation will cause errors in the output. If V_{DD} and other operating conditions remain fixed, the error generated by a computational unit due to VOS will retain information about the process variation. Since process variation is a random process, by profiling this error, one can distinguish the computational unit and generate a unique device signature for the circuit.

6.4.1 Errors in an Approximated Circuit

To understand the effect of process variation in voltage over-scaling, we have studied the error profiles generated by adders. Venkatesan et al. have provided process variation independent error profiles for Ripple Carry Adders (RCA), Carry Look-ahead Adders (CLA), and Han–Carlson Adder (HCA) [16]. It was found that the error probability increases as the number of critical paths that fail to meet the timing constraint increases [16]. Therefore, with the presence of randomness in the manufacturing process, the variations in the transistors in the critical paths will have a significant contribution to the errors produced by the adders.

Among the adders, it was found that Han–Carlson adder fares the poorest regarding producing correct result with voltage over-scaling. RCA performs better than HCA and the probability of error increases slowly with the length of the adder [16]. CLA performs best among these three adders. If one wants to extract fabrication variation related information, she needs to be careful on the choice of circuits. For example, HCA can suppress the variation dependent errors with the errors due to scaling effects, and CLA can repress impact of process variation due to its timing forgiving nature. Therefore, for our discussions, we have used RCA that has the potential to preserve process variation related artifacts.

6.4.2 Error Modeling

If a given circuit is operated with a clock period that is less than the maximum delay produced by the circuit, then the circuit output becomes a function of current and previous input values [16]. Therefore, the output data in a circuit under VOS not only is a function of process variations but also a function of the input values applied to the circuit. Hence, for a combinational adder with two operands $\mathbf{x}$ and $\mathbf{y}$, we can write the current output $\mathbf{z_i}$ of a voltage over-scaled adder as a function of current inputs $\mathbf{x_i}$, $\mathbf{y_i}$, and previous inputs $\mathbf{x_{i-1}}$, $\mathbf{y_{i-1}}$. Therefore,

$$\mathbf{z_i} = f(\mathbf{x_i}, \mathbf{y_i}, \mathbf{x_{i-1}}, \mathbf{y_{i-1}}) \tag{6.3}$$

where $f()$ defines a process variation dependent addition. This dependence on previous inputs can cause cascading errors in the output. Therefore, to correctly predict the output of a voltage over-scaled adder, one (say Alice) can take the following measures.

1. Save the output data for the set of input patterns that will be used on that circuit. For example, if only a set S_I containing n-input pattern will be used for processing in a given adder, then Alice needs to save outputs for all the combination of the pattern resulting from S_I. This would reveal the partial behavior of the circuit for a subset of input data.
2. Profile the adder for all possible input patterns. For profiling the adder, one not only needs to consider all possible current input values but also requires previous input values. Therefore, a correct profile of an n-bit voltage over-scaled adder would consist a table of entries comprised of all possible current input value times all possible input values in the previous step. This would amount to $2^{2n} \times 2^{2n} = 2^{4n}$ entries of the input values and the corresponding output values. Since all the additions are not incorrect and dependent on the previous values, one can significantly reduce the size of the profile by storing only the cases where the adders provide inaccurate results.
3. Use a delay-based graphical model to learn the properties of the adder. Since Alice has the adder, she can profile the device and use this profile to create a conditional probability table for a Bayesian network.

   Our construct of using process dependent delay information for generating responses to random challenges requires the verifier have access to the exact model of the device. Given the access to a large number of input-output of the adder, a verifier can build a model of the adder using Probably Approximately Correct (PAC) learning model in polynomial time. We can prove that such constructs are properly learn-able using the proof presented by Ganji et al. [17].

6.4.3 Assumptions

From the discussions above, we are stating the following requirements to design authentication protocols using a voltage over-scaling based VLSI system. The computing unit (i.e., the adder) must fulfill these requirements for ensuring a secure authentication.

M1 Voltage over-scaling can produce process variation dependent errors in the computing unit.

M2 Errors produced due to voltage over-scaling are not random noise but reproducible information since they merely reveal the output of the circuit at a lower operating voltage.

M3 If one has access to the input and output ports of this circuit, he can adequately model the behavior of the computation performed by this unit.

M4 Such model discussed in R3 would be unique for each computational unit since the manufacturing-dependent process variations are random in nature.

To understand the effects of voltage-over-scaling on processed data, we have simulated simple image processing example using RCA to analyze that effect of process variations, voltage variations, and temperature. We have performed our simulations in HSpice platform using the FreePDK 45nm libraries [18]. To introduce process variation in our design, we have used 200 modified NMOS and PMOS models with variable threshold voltages. Gaussian distribution with a ±7.5% standard variation is assumed for the variation of the threshold voltages. This modified NMOS and PMOS transistor models are randomly chosen to build 100 different versions of each standard cell in the FreePDK 45nm library. We have designed our digital circuits in Verilog and synthesized them using the Cadence Virtuoso RT compiler. The synthesized design is then converted into an HSpice netlist with standard cells randomly chosen from our modified library.

We present a simple image processing application based on the superimposition of two images under general operating conditions and compare their results. The image processing application superimposition reads the 8-bit values stored at every pixel location for any two given image and add the values (an example of superimposition is given in Fig. 6.2). The processing is first done on an accurate adder and then on two voltage over-scaled ripple carry adders (as shown in Fig. 6.1) with process variations.

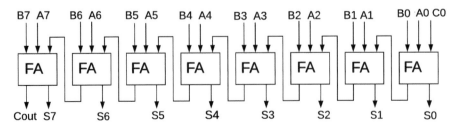

Fig. 6.1 8-bit-ripple carry adder

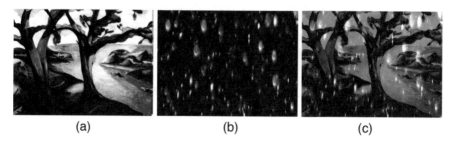

Fig. 6.2 An example of superimposing two images. We have used two grayscale images (**a**) trees, and (**b**) snowflakes from MATLAB library to generate the superimposed image, (**c**) snowfall

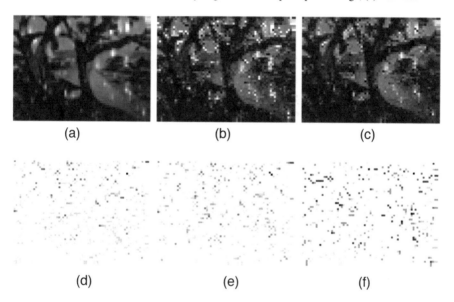

Fig. 6.3 An example of the effect of process variations in voltage over-scaling based computation. In (**a**) the grayscale image Snowfall is computed using trees and snowflakes without voltage over-scaling using $v_{dd} = 1V$; in (**b**) and (**c**) the image is computed under voltage over-scaling using $v_{dd} = 0.4V$ with two adders which are identical in every aspect, except the process variation of the transistors; (**d**) and (**e**) shows the error pattern found in the figure (**b**) and (**c**). This error pattern shows the deviations for each adder from the correct image. Subfigure (**f**) shows the difference between the two error patterns (**d**) and (**e**). The source images were downsized to 52×40 pixels for reducing computation time

From visual observation of Figs. 6.3a, b, and c, one can clearly notice the effect of voltage over-scaling in this simple image processing application. Figure 6.3b and c are somewhat distinguishable if an observer pays close attention. We generate the error patterns by calculating the difference between each pixel value in the approximated result and the correct result. Although it is difficult to distinguish the difference between the error patterns shown in Fig. 6.3d and e, if we plot their differences as done in Fig. 6.3f, one can notice the effect of process variation in the

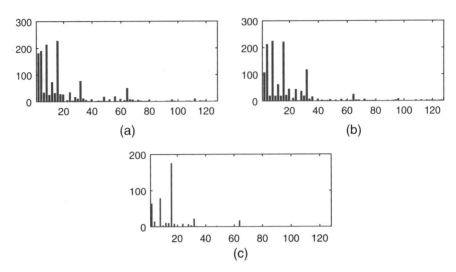

Fig. 6.4 A histogram of the Euclidian distances (**a**) between the Figs. 6.3a and b; (**b**) between the Figs. 6.3a and c; between the Figs. 6.3b and c. The x-axis in the sub-figures represents bins of all the possible distance values from 1-128, and the y-axis in the sub-figures represents the number of pixels (in the two corresponding computation of Snowfall images) that share the same distance values

approximately computed result. If we plot the histogram of the Euclidean distances between the Figs. 6.3a, b and Figs. 6.3a,c (as done in Fig. 6.4), then we can see some interesting results on the error pattern. It can be noted that the peaks of this histogram mainly appear at 2,4,8,16,32,64..., which suggests that most of these errors in the approximated results come from a single bit-errors. Furthermore, the peaks at 1,2,4,8,16 are higher in most cases, revealing the fact that for these two adders most of the errors are in the LSBs.

6.4.4 Authentication Protocol Using Approximation Errors

For single entity authentication, we will assume an interactive protocol VOLtA [19] between a single prover $\mathcal{P}$ and a verifier $\mathcal{V}$. Both the prover and the verifier have some knowledge about a shared secret **x**. The secret is generate through a key generation procedure $KeyGenVOLtA(1^\lambda)$, where λ is a security parameter. The authentication protocol responds with the outputs accept or reject after a successful run of the protocol.

Assumptions Assuming that one has access to hardware that satisfies the requirements M1-M4 presented in Sect. 6.4.3. This authentication protocol assumes that the prover has a voltage over-scaled computation unit (H) that generates process

Table 6.2 Single round interactive authentication VOLtA

Prover($\mathbf{x_1}, \mathbf{x_2}, H$)	Verifier($M, \mathbf{x_1}, \mathbf{x_2}, \epsilon$)
	$\mathbf{R} \overset{\$}{\leftarrow} \mathbb{Z}_p^{\ell \times n}$
$\overset{\mathbf{R}}{\leftarrow}$	
Calculate $\mathbf{L} = H(\mathbf{R}, \mathbf{x_1}) = \mathbf{R} + \mathbf{x_1}$ using the adder and then calculate $\mathbf{z} = \mathbf{L} \oplus \mathbf{x_2} = (\mathbf{R} + \mathbf{x_1}) \oplus \mathbf{x_2}$	
$\overset{\mathbf{z}}{\rightarrow}$	
	Calculate $\mathbf{z'} = M(\mathbf{R}, \mathbf{x_1}) \oplus x_2$. If distance $(\mathbf{z'}, \mathbf{z}) \leq \epsilon$ accept

dependent errors. The verifier either knows the correct model or profile (M) to simulate the computation unit.

Algorithm 6.2 RNG based key generation for voltage over-scaling based authentication

1: **procedure** $(\mathbf{x_1}, \mathbf{x_2}) \leftarrow KeyGenVOLtA(1^\lambda)$
2:    Sample $\mathbf{x_1} \overset{\$}{\leftarrow} \mathbb{Z}_p^\ell$
3:    Sample $\mathbf{x_2} \overset{\$}{\leftarrow} \mathbb{Z}_p^\ell$
4:    **return** $\mathbf{x_1}, \mathbf{x_2}$

Enrollment The prover and the verifier use the key generation procedure $KeyGenVOLtA$ to generate secrets $\mathbf{x_1}, \mathbf{x_2}$. ϵ is the predetermined error threshold for the authentication.

Verification Single round of verification for this protocol is provided in Table 6.2.

Note that, the distance in the protocol can be measured by standard distance measurement functions such as Hamming distance or Euclidean distance. Also, with multiple keys, the verifier can authenticate numerous users using the same device. Moreover, the verifier can verify the prover over different devices if the verifier knows the correct model of those devices.

6.4.5 Evaluation of the Protocol

For our discussions on threat models and attacks, we assume that Alice tries to authenticate Bob over an untrusted channel where Malice performs the following attacks to obtain the security keys or being erroneously recognized as Bob. Below we discuss the potential attacks on VOLtA.

This is a simple two-factor authentication scheme, which requires that a shared secret is known (i.e., key $\mathbf{x_1}, \mathbf{x_2}$) and, another shared secret possessed (i.e., properties of the voltage over-scaled adder). To prove the effectiveness of our proposed VOLtA, we start from analyzing the potential weakness, for the case when we have a perfect adder. If the adder is perfect when calculating $\mathbf{L}$, this protocol is not secure. Assume the following scenario: the malicious attacker Malice is pretending to be Alice, and she wants to resolve Bob's key $\mathbf{x_1}, \mathbf{x_2}$ by sending some messages $\mathbf{R}$ and receiving the corresponding $\mathbf{z}$ from Bob. Then she will apply eavesdropping and bit manipulation techniques to recover the key.

However, when applying the voltage over-scaling approach, the addition will become non-deterministic because the physical variations will affect the arithmetic result as discussed in Sect. 6.4.1. Therefore, the result of $\mathbf{M}(\mathbf{R}, \mathbf{x_1})$ cannot be accurately predicted. As a result, the bitwise attacking scenario fails. Overall, for the security of this protocol, the uncertainty of the calculation in the arithmetic function needs to be guaranteed.

Random Guessing Attack The most straightforward attack that an adversary can perform on the authentication protocol is to try and randomly guess the authentication keys. To accomplish this attack, Malice tries to imitate Bob and responds to Alice's query with a random guess $\mathbf{z'}$.

The security of VOLtA is tied not only to the key $\mathbf{x_1}, \mathbf{x_2}$ but also the property of the approximate adder. Since Malice neither has information about the key nor about the hardware properties, the success rate of such attacks exponentially decreases with the increase in the number of bits in the security keys and with the increase in the uncertainty of the results produced by the adder.

Eavesdropping Attack For eavesdropping attack, Malice eavesdrops on some communications between Alice and Bob and records Bob's response to each challenge from Alice. Later, for a known query of Alice, Malice can answer correctly using her records.

Alice sending random string $\mathbf{R}$ each time can easily thwart such attacks because in that scenario response calculated by Bob will be different for different cases. Since Alice knows the correct model of Bob's circuit, she can send random string every time for authentication. Therefore, VOLtA would be useful in counteracting such attacks.

Man-in-the-Middle Attack Man-in-the-middle attack constitutes the case where Malice pretends as Alice and communicates with Bob. Malice sends random authentication strings to Bob and collects his response. This attack would be difficult to perform if Bob has some knowledge about the input sent by Alice. But this would violate the requirement of randomized string to prevent other attacks. However, our authentication mechanism will succumb to a MITM attack that queries Bob with a correct challenge, learn it, and give Alice the response.

Compromised Key One of the most active attacks on these protocols is the situation where the key $\mathbf{x_1}, \mathbf{x_2}$ and the model M are leaked to an attacker. Since this case breaks the fundamental Kerckhoffs's principle, both the protocols will fail in the face of such attacks.

Therefore, encryption techniques need to be applied to protect Alice's database to ensure the security of the keys and models. It should be noted that this would provide security in the case when the key K is compromised because it is a two-factor authentication protocol where the property of the adder is unknown to the attacker. Therefore, without having the model of the device, the attacker would still have to resort to random guessing or other attacks to resolve a correct response.

Learning-Based Attack This attack is a combination of eavesdropping attack and learning attacks. Malice eavesdrops on the communication during authentication and with the challenge and response records, Malice models the voltage over-scaled approximate adder using a learning model. Thus, Malice can create a delay-based graphical model for the adder with partial observables. Malice can estimate a conditional probability table, and the chance of success in getting the model for the adder will increase the number of trials that Malice can perform. However, the output of the adder is XORed with x_2, and therefore such attacks will be difficult to perform without the knowledge about x_2.

Side Channel Attack One of the most common side channel attacking techniques for encryption is static or differential power analysis. Researchers have shown that the power analysis can severely reduce the security of Ring Oscillator PUF, which is a well-known secure primitive for key storage and authentication. However, our proposed voltage over-scaling is more resistant to side channel attack because the arithmetic units are working under much lower voltage. As we mentioned before, the dynamic power consumed in a VLSI chip is squarely proportional to the supply voltage, the power consumption during authentication process will be very low, making it difficult to capture accurate power consumption. Since the adder does not generate correct result, even though the attacker can measure the exact power consumption, he cannot apply the model of an accurate adder for regression. The real model M is hidden in the process variations.

A detailed discussion on the experimental validation of VOLtA can be found in [19]. In summary, in this section, we introduce a voltage over-scaling based authentication scheme(VOLtA) that uses the random physical variation of a VLSI system. VOLtA profiles the hardware used for computation in a reduced voltage operation and uses the underlying hardware fingerprint for authentication purposes. This authentication protocol requires no additional hardware on the claimant side to implement. This lightweight protocol can be useful in IoT applications where the interconnected Things face extreme power, cost, and area constraints.

6.5 Authentication Using Memory Components

In this section, we address the problem of secure and lightweight authentication of single and multiple entities using memory system components. We assume an entity (say Alice the verifier) tries to simultaneously authenticate k-out-of-n-other entities

$(B_1, \ldots B_k)$. This problem can be simplified as simultaneous verification of k-out-of-n cryptographic keys. If these keys are generated from a secret that Alice knows or possesses, then, instead of checking n-keys separately, Alice can use the keys provided by $B_1, \ldots B_k$ to regenerate her secret and authenticate all entities at once. Note that, as Alice uses a k-out-of-n secret sharing scheme, Alice can reconstruct the secret from any of the k-users shares. The problem of authenticating a single user (say Bob) then reduces to a 1-out-of-1 authentication problem. For avoiding collusion among the users, this procedure requires Alice to have some interference in key/password generation process during the registration of the users.

We consider emerging resistive random access memory (RRAM), also known as memristors, devices for secret sharing based authentication. Memristor-based crossbars are used for designing one of the first memristor-based PUFs called nano-PPUF [20]. A simulation model of the physical design of a public PUF is publicly available; however, simulation complexity can create a time-bound authentication protocol. The attack model assumes a computationally bounded adversary unable to simulate the exact output for a given PUF design. The non-linear equations governing the current–voltage relationship of memristors and the viability of fabricating large memristive crossbars provide the simulation complexity required by this PUF model.

In this PUF design, a public registry contains the simulation model for a given user's (Bob's) memristive PUF. When Alice wants to authenticate Bob, she first sends a random challenge vector $\mathbf{V_C} = \{v_1, v_2, \ldots v_n\}$, where, v_i represents a physical input. For the given PPUF at [20], $\mathbf{V_C}$ is the voltage applied to an $n \times n$ memristor-crossbar. Since Bob has the physical memristor, he can correctly respond to Alice's challenge. He sends the correct response vector $\mathbf{V_R}$. For a computationally bounded attacker Malice, completing this step would require simulating the complete crossbar, which would be computationally prohibitive. For completing the authentication, Alice then picks a subsection of Bob's crossbar (a polyomino) and requests the voltages at the boundaries of this polyomino. Bob sends the measurement and simulation results. Alice can accurately simulate the smaller polyomino using $\mathbf{V_C}$ and $\mathbf{V_R}$, and verify Bob's results. Thus, Alice can authenticate Bob.

This initial PUF design suffers from several crucial drawbacks. The crossbar simulation and the results from the physical crossbar will only match if the physical conditions that affect the current in a memristor (such as temperature, history of current flow, aging) remain the same. This is a difficult condition to fulfill for such design. Moreover, an attacker can try machine learning and model building attacks on passively obtained challenge responses to breaking the authentication scheme. Additional improvements considering these physical effects on this PUF design are discussed in [21] and [22].

6.5.1   Requirements and Utility Functions

Before proposing memristor-based entity authentication protocols, we first describe
the basic memristor utility functions that are critical for these protocols.

RESET: A RESET operation uses a fast negative biased pulse between the top and
the bottom electrode to put a memristor to the high resistive state (HRS).

SET: A SET operation puts a memristor to the low resistive state (LRS). For
authentication purpose, we will use the bias dependent write time variability
during set (transition from HRS to LRS) operation. We will use multiple short
duration ON pulses for setting the device instead of a longer ON pulse.

SET Pulse Count (SPC): The number of ON pulses required to transit a memristor
from HRS to LRS.

Precondition: A k-bit precondition operation applies k consecutive ON pulses to the
memristor device. If the device reaches the LRS after k_n pulse where $(k_n < k)$,
the device is RESET and the remaining ($i.e.$, $(k - k_n)$) pulses are applied.

Read State: No pulse for one cycle. Then detect whether the memristor device is at
HRS or LRS.

From common device features of memristors, we specify our assumptions on the
expected device behavior that is needed for the design of the secret sharing based
authentication protocols, and also give the justification of these requirements and
assumptions. For our discussion, let us assume that the SET operation is achieved
by singular or multiple constant duration voltage pulses which are denoted by ON
pulses.

R1 Changes due to consecutive ON pulses are *additive* in nature.
    When given sufficiently high frequency and proper amplitude, short duration
    ON pulses can cause memristor resistance state transitions just like a longer
    single voltage pulse. For example, if the device is not already in LRS, two
    consecutive ON pulses would put the device in a lower resistive state than a
    single ON pulse.

R2 The memory elements are *monotonic*.
    If the device is not in an LRS, applying an ON pulse will always decrease the
    resistance, and its effect cannot be undone without resetting the device.

R3 Intermediate resistive state transitions are random; however, the number of ON
    pulses remains nearly the same for different programming cycles (Fig. 6.5).

R4 The number of short duration pulses during the state transition cannot be
    predetermined without information about the programming conditions.
    This is the discrete version of requirements R1 under the assumption of R3.
    In other words, one cannot imply how many short duration pulses have been
    applied by observing the intermediate resistance value during the state transi-
    tion. Furthermore, to learn the exact/an approximate number of pulses required
    for a given state transition at a given bias, one memristor to observe a complete
    programming cycle. At a given bias, it would be impossible to extrapolate the

Fig. 6.5 For memristive device with different programming cycle different voltages are required to reach the same resistive level [23]

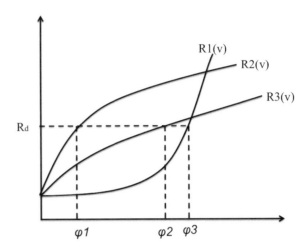

number of pulses required for a complete transition by measuring the resistance change of the device for a small period.

R5   There exists a higher current level for which the device can be RESET to a non-recoverable high resistive state (NRHRS), i.e., the device cannot be SET again after it reaches the NRHRS.

R6   Memristive devices stay at random resistive states after fabrication. A forming bias voltage V_{form} is required to make the devices programmable.

6.5.2   An Example of Single Entity Authentication: Protocol I

For single entity authentication, we will assume an interactive protocol between a single prover $\mathcal{P}$ and a verifier $\mathcal{V}$. Both the prover and the verifier have some knowledge about a shared secret **x**. The secret is generate through a key generation procedure $KeyGen(1^\lambda)$, where λ is a security parameter. The authentication protocol responds with the outputs **accept** or **reject** after a successful run of the protocol. First, we present a simple key generation procedure using the intermediate state transition of a memristor.

In this protocol, the prover and the verifier participate on one time enrollment phase that generates the secret using the key generation algorithm $KeyGen01$ and distribute to the parties. For authentication, multiple round of interactive authentication (as presented in Table 6.3) is performed.

Assumptions Assume that there exists high variability in the fabrication of the devices and the state transition of a memristor with small voltage pulses is unique for each device and highly repeatable.

Algorithm 6.3 Key generation using resistive states

1: **procedure** x ← $KeyGen01(1^\lambda, \alpha)$ $\triangleright \alpha \in \mathbb{N}$ is the experiment repetition value
2:     For a given memristor R, select a set of $\{V_{select}, V_{BL}, V_{WL}\}$ that demonstrates a complete programming cycle with an average write time $t_{wr,avg}$.
3:     Assume the key length $\ell = nb$, where $n, b \in \mathbb{N}$. $\ell, t_{wr,avg}, V_{select}, V_{BL}, V_{WL}$ values depend on the security parameter k.
4:     **for** j=1 . . . α **do**
5:         **RESET** R.
6:         Apply n-consecutive pulses (where $n = \ell/b$, and b is the number of bits used for representing a floating point number in the system) at the word_line with pulse width $t_p = t_{wr,avg}/n$ and measure the corresponding current I_i through R after ith pulse, where $i \in \{1, \ldots, n\}$.
7:         $\mathbf{X}[j, :] \leftarrow I_1 || I_2 \ldots, || I_n$.
8:     $\mathbf{x_a}[i] \leftarrow$ majority values of the ith column of **X**
9:     Use a b-bit ADC to convert $\mathbf{x_a}$ to a binary string **x**.
10:     **return** x

Enrollment The prover and the verifier use a random memristor M and generate the ℓ-bit secret $\mathbf{x} \leftarrow KeyGen01(1^\lambda)$. They also share a b-bit precondition value s. The authentication protocol assumes that the verifier has the memristor and the prover knows the secret state transitions of the memristor given by **x**. ϵ is the predetermined error threshold for the authentication.

Table 6.3 Single round interactive authentication for Protocol I

Prover($\mathbf{x}, s$)	Verifier(M, s, ϵ)
	– **RESET** M
– Find the current I_i for the precondition value s using **x**	– **Precondition**(M,s)
	$- c \xleftarrow{\$} \{i \in \{1, 2, \ldots n - s\} : n = \ell/b\}$
	– **Precondition**(M,c)
	– Measure the current I_{rand} with the parameters $\{V_{select}, V_{BL}, V_{WL}\}$ specified at the key generation step
	$\xleftarrow{I_{rand}}$
– Find the distance (r) between the current values I_i and I'_{rand} at **x**, where I'_{rand} is the near-most value of I_{rand}. r is approximately the number of ON pulses required to be applied to memristor R at the s-state to produce I_{rand}	
$\xrightarrow{r}$	
	– If $abs(r - c) < \epsilon$ accept, otherwise reject

Update After one successful round of authentication the prover and the verifier both update $s \leftarrow s + r$.

Protocol I requires 1T1M memristive cell with current measuring capabilities. This protocol is designed based on the assumption that reducing the bias voltage of the memristors increases the total write time and enables multi-level operation of the device that is less susceptible to the effect of environmental variation and Joule heating. Although elongating the write time of a memristor memory cell is less favorable for memory operation, it is a welcoming feature in security, since it increases the read-out time for the entire contents of the memory in several orders of magnitude.

For the realization of this protocol, we assume that the I-V curve during state transitions is repeatable for a memristor for a given bias and environmental condition. The non-linearity between the applied voltage and the corresponding resistance can create unique state transitions for a given memristor. When the resistive characteristics are different for two different memristors, different amount of voltage is required for obtaining the same level of resistance [23]. Quantifying the application of bias voltage into voltage pulses, one can generate secrets based on the number of pulses needed for a given state transition.

6.5.3 Security Analysis of Protocol I

We assume an attacker $\mathcal{A}$ who wants to generate an **accept** from the verifier $\mathcal{V}$.

Theorem 6.1 *If $\mathcal{A}$ guesses at random, she has $(1/n)$ chance of success, where $n = f(\lambda) = \ell/b$ is the number of voltage pulses required for a complete SET-RESET cycle.*

Proof Since, the memristor requires n pulses for a complete SET-RESET cycle, the response r is in $\mathbf{r} = \{0, \ldots n\}$. Therefore, the chance of success for random guessing is $1/|R| = 1/n$

The security can be improved by introducing multiple rounds and for β rounds the chance of success for an attacker reduces to $(1/n^{\beta})$.

Theorem 6.2 *Assume $\mathcal{A}$ can eavesdrop on the challenge-response pair (I_{rand}, r) for the authentication. Then she can learn the complete challenge response by eavesdropping on an average of $O(n^2)$ consecutive communications.*

Proof If n-pulses are used for complete programming of the device, then, we can assume that there are n distinct states in the device. Transitions between the states can be represented by a fully connected graph of n vertices with $n(n-1)/2$ edges. Therefore, to completely learn the state transition $\mathcal{A}$ needs to solve an average of $O(n(n-1)/2) = O(n^2)$ equations describing consecutive runs for the authentication protocol.

Therefore, we can see that the protocol is weak against eavesdropping attack. Lowering the bias voltage of the memristor can exponentially increase the number of pulses (n) and thus increase the security against eavesdropping and random guessing attack; however, it would create increasing storage burden of $O(n)$ to the prover. Furthermore, this protocol requires the extensive capability of current measurement. This problem can be addressed by using differential measurement techniques where the resistance of the memristor is compared with other fixed resistors with different resistances.

Observation 6.1 (Secure Storage) Access to the memristors without the knowledge of biasing condition will reveal μ-knowledge to an attacker, where μ depends on the number of parameters required to correctly model operating cycles of a given memristor.

In this protocol, we keep the memristors response to the input pulse (the state of the memristor p_i) as the secret shared between the prover and the verifier. Based on device requirement R1 and R4, the attacker with single access may be able to obtain whether a memristor is at HRS or LRS, but he will not be able to find the cycle-accurate state information of the memristor. He will need to access the memristor an average of $O(\mu)$ times (assuming a linear model) to create the programming model of the device.

An essential drawback of this protocol is that it requires the device to have same programming cycles, *i.e.*, the device must maintain the similar programming behavior over different cycles. For a reliable application, we need to either reduce such restrictions or provide proper error-correction mechanisms such as majority voting and fuzzy extractor algorithms [24] for a secure application. Helper data in error correction can leak information, and therefore, such construction must be carefully designed.

6.5.4 Multiple User Authentication

To address the problems regarding multi-user authentication, we propose to design a Visual Cryptography inspired memristor-based multi-user authentication scheme. A small motivational example for 2-out-of-3-user authentication using memristor-based threshold detector circuit is given below to illustrate the key concepts.

Assume that an entity Alice wants to simultaneously authenticate three users B1, B2, and B3. Alice chose a random key K, and for each bit of the key, for each user, Alice randomly chooses a row from the following matrix C_0 and C_1 using the given rules and gives the corresponding 3-bits to the users. The matrices are given as [7]:

$$C_0 = \{\text{all the matrices obtained by permuting the columns of} \begin{bmatrix} 1 & 1 & 0 \\ 1 & 1 & 0 \\ 1 & 1 & 0 \end{bmatrix}\}$$

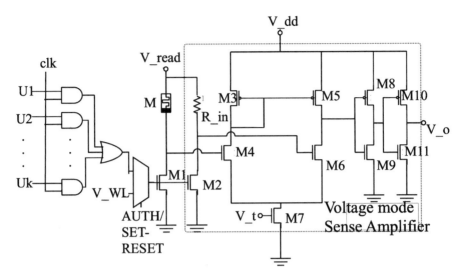

Fig. 6.6 Circuit for applying the pulsed input from k-users and the assisting voltage-mode sense amplifier-buffer circuit for reconstructing the secret

$$C_1 = \{\text{all the matrices obtained by permuting the columns of } \begin{bmatrix} 1 & 1 & 0 \\ 0 & 1 & 1 \\ 1 & 0 & 1 \end{bmatrix}\}$$

and the distribution rules are:

  i If the ith-bit of the key is 0, distribute the rows of C_0 to B1, B2, and B3.

  ii If the ith-bit of the key is 1, distribute the rows of C_1 to B1, B2, and B3.

Note that for a single user, it would be impossible to guess the key bit. For authentication, Alice lets any two or more users access the memristor for the duration of three pulses. The users apply their keys to each pulse. The keys are taken through an OR gate as shown in Fig. 6.6. An ON pulse is used at a cycle if the result of the OR gate is one. Otherwise, no pulse is applied. If the ith-bit of the key is zero, then for only two of the cycles the users will give an ON pulse, but there will be no ON pulse for one cycle. However, if the ith-bit is 1, there will be three ON pulses. As a result, after the duration of three pulses, the device's resistance should be lower if the key was 1 and higher if the key was 0.

A threshold detector can easily detect this and reconstruct the key. This scheme can easily be modified to an n-out-n or a k-out-of-n authentication system. One weakness of the above example is that to reconstruct a bit of the key; the users do not necessarily need Alice. If any two of the valid users come together, they can restore the bits. If we want a hardware dependent authentication, where the participating device should also be a factor during authentication, this might cause a collusion problem. However, this issue does not exist if Alice has some interference in key generation.

Table 6.4 Single round interactive authentication for multiple prover, single verifier

Provers($\mathbf{S_1}, \mathbf{S_2}, \ldots, \mathbf{S_k}$)	Verifier($M, \mathbf{x}$)
	–**RESET** all memristors of M
	– Set the bias point such that all memristors require p-ON pulses to move to the LRS, where p is the number of columns of $\mathbf{B_0}, \mathbf{B_1}$
for all $i \in \{1, 2, \ldots \ell\}$	
for all $j \in \{1, 2, \ldots p\}$	
Simultaneously apply all $S_*[i, j]$ on the U_* gates at Fig. 6.6	
	– If $\mathbf{V_{out}} = \mathbf{x}$, then accept, else reject

---

Algorithm 6.4 Key generation and distribution for multiple user authentication

---

1: **procedure** $\mathbf{x} \leftarrow KeyGenM(1^\lambda)$
2:      Sample $\mathbf{x} \xleftarrow{\$} \mathbb{Z}_2^\ell$
3:      **return** $\mathbf{x}$
4: **procedure** $(\mathbf{S_1}, \mathbf{S_2}, \ldots, \mathbf{S_k}) \leftarrow KeyDistM(\mathbf{x})$
5:      Consider a ground set G consisting of k elements $g_1, g_2, \ldots g_k$. The subsets of G with even cardinality are $e_1, e_2, \ldots, e_{2^{k-1}}$ and odd cardinality are $q_1, q_2, \ldots, q_{2^{k-1}}$
6:      Construct two Boolean matrices $\mathbf{B_0}$ and $\mathbf{B_1}$ of size $k \times 2^{k-1}$ such that, $\mathbf{B_0[i, j]} = 1$ iff $g_i \in e_j$ and $B_1[\mathbf{i, j}] = 1$ iff $g_i \in q_j$ for $i \in \{1, \ldots k\}$ and $j \in \{1 \ldots 2^{k-1}\}$
7:      Construct collections ξ_0 and ξ_1 by permuting all the column of $\mathbf{B_0}$ and $\mathbf{B_1}$
8:      **for all** $i \in \{1, \ldots, \ell\}$ **do**
9:          $\mathbf{C_0} \xleftarrow{\$} \xi_0, \mathbf{C_1} \xleftarrow{\$} \xi_1$
10:         **for all** $j \in \{1, \ldots, k\}$ **do**
11:             **if** $\mathbf{x}[i] = 0$ **then** $\mathbf{S_j}[i, :] = \mathbf{C_0}[j, :]$ ▷ distribute the rows of C_0.
12:             **else** $\mathbf{S_j}[i, :] = \mathbf{C_1}[j, :]$ ▷ distribute the rows of C_1.

For multiple user authentications, we will assume an interactive protocol between a multiple provers $\mathcal{B}$ and a verifier $\mathcal{V}$. All the provers and the verifier have some knowledge about a shared secret $\mathbf{x}$. The secret is generated through a key generation procedure $KeyGenM(1^\lambda)$, where λ is a security parameter. The authentication protocol responds with the outputs accept or reject after a successful run of the protocol. Here we formally describe the simple key generation and distribution procedure which is similar to the one proposed by Naor et al. [7].

Assumptions For the authentication scheme described below, we assume that R1-R6 described at Sect. 6.5.1 holds. We also expect that the verifier owns the authentication hardware and controls the bit-line voltage (V_{BL}). The verifier also acts as the dealer $\mathcal{D}$ that knows $\mathbf{p}$ (where ($p[i]$) is the number of ON pulses required for pushing the resistance of the memristor $M[i]$ from a fixed HRS state to the reference resistance (R_{in}).

Due to the fabrication variations, each memristor would have different write time, which will lead to different (but of the same order) values of p_i for various devices at

---

Algorithm 6.5 Extraction of correct precondition values

1: **procedure y** ← $ErrCorr(k, M)$ ▷ M is a memristive array of length ℓ
2:    **for all** $i \in \{1, \ldots, \ell\}$ **do**
3:        **RESET** $M[i]$
4:        Apply k ON pulses on M[i]
5:        **READ** M[i].
6:        **if** $M[i] = HRS$ **then**
7:            Apply t-pulses to put $M[i]$ in LRS
8:            $y[i] = t$
9:        **else** $y[i] = 0$
10:   **return y**

---

a given V_{BL}. Since the dealer/verifier owns the equipment, she knows the value of p_i for any given Ri. As different memristor requires different numbers of ON pulses, additional 1s need to be padded in each row of C_0 and C_1. A more straightforward solution to this problem is to precondition each memristor with this additional 1s. Before each round, the verifier can precondition each memristor with precondition value **y** so that each requires the same number of extra ON pulses to reach LRS as discussed in procedure ErrCorr. Another way is to use the block-length defined constructs. Block length is the number of ones resulted by OR-ing all the columns of C_1. Therefore, the dealer can design block-length adjusted matrices C_0' and C_1' depending on the number of ON pulses required for a state transition and follow $KeyDistM$ to share a bit ($\mathbf{x}_i$) of the key **x** (Table 6.4).

To make a hardware dependent authentication scheme, the dealer can choose smaller block length or pad less number of 1s at the end of C_1' for some of the random bits of the key X. Since the provers do not have access to the hardware, they do not know the exact value of p_i's for a given $M[i]$. Therefore, the dealer can share a 0 by sharing contents of C_1' for cases where C_1' is not correctly generated from C_1. Thus, it would be impossible for k-users (or even all the users) to collude and guess the secret key.

For developing a memristor-based secret reconstruction and authentication circuit, we need to extract the current state of a given device. It should be noted that improper reading of a memristor cell with high bias voltage across the cell can change its states, and therefore, the read operation should be done carefully. To accomplish this, we have used a simple CMOS compatible differential sense amplifier. The resistance of the device R_x is compared with an on-chip resistance R_{in} that has a threshold resistance value. If $R_x < R_{in}$ the circuit outputs 1, else it outputs a zero. The basic circuit is shown in Fig. 6.6 which can be used for reconstructing the shared secret.

6.5.5 Security Analysis of Multi-User Authentication Protocol

A weak adversarial model can be assumed from a group of dishonest participants in a k-out-of-n secret sharing model. Such attack is defined as cheating where a group of unscrupulous users uses fake shares to alter the secret that was distributed at the initial phase. This is a typical attack scenario in visual cryptography-based secret sharing, and there are several practical countermeasures proposed by Horng et al. [25]. Since we have used visual cryptography for authentication purposes with the secret shared at the initialization phase, cooperation between multiple dishonest users would create a denial-of-service attack on the authentication system. If the dealer uses hardware dependent scheme, fraudulent users cannot create a cheating attack or reconstruct the secret by themselves since the dealer/verifier owns the hardware. Additionally, for detecting such attacks, the dealer can put redundant information in each share that constructs a 2-out-of-2 secret sharing with dealer/verifier and the prover.

In summary, non-volatile memory based devices and circuits are monotonic. Exploiting this monotonicity can be useful in designing secure circuits and security protocols. In this work, we have connected the "additive" and monotonic nature of memristor devices with secret sharing based user authentication ideas. We have reported the designed protocol and the necessary circuits required for single and multiple user authentications using memristor-based hardware. These robust, hardware dependent user authentication schemes can be useful in developing security primitives in the extremely resource-constrained IoT applications.

6.6 Authentication and Spoofing Detection Using Hardware Clocks

The progress toward the Internet of Things (IoT) is highly dependent on the secure and successful integration of a trusted and robust geospatial localization and clock synchronization mechanism for *Things* across a large distributed network. Currently, both these functions are predominantly provided by the Global Positioning System (GPS). Recently, there have been several demonstrations of weaknesses and vulnerabilities of GPS signals and GPS receivers [26–30]. Most of the analysis on GPS spoofing is directed toward the spoofing of position data. However, GPS system is also used for large area clock synchronization, and therefore, attacks on GPS signals can impact networked infrastructure where accurate time keeping is essential.

In this section, we present a data-level authentication mechanism for GNSS signals that rely on intrinsic hardware properties of a free-running crystal oscillator. Since the free-running oscillator is located on the device and not externally synchronized, it presents a minimal attack surface while exhibiting a strong correlation with authentic GPS signals. We propose that "*anomalies*" in the correlation index can authenticate received GPS data. Our approach is simple, fast, and can perform in near real-time. Additionally, the design is low cost and can act as an add-on to virtually any GPS receiver.

6.6.1 The GPS System

The GPS system consists of satellite transmitters and (usually) terrestrial receivers. Each transmitter satellite broadcasts at two frequencies: 1575.42 MHz (L1) and 1227.6 MHz (L2). The L1 carrier messages are available for civilian purposes. These messages are not encrypted but modulated with pseudo-random noise (PRN) codes to distinguish each satellite. The L2 carrier is modulated by encrypted codes and reserved for military purposes. Message from each GPS satellite contains information about the position of the satellite and the time of the onboard atomic clock [31].

To calculate true receiver-to-satellite distance, the receiver requires the range (r_{true}) of a satellite at a given time. This can be calculated by multiplying the signal propagation time (from the transmitter to the receiver) with the speed of light (c). Then, for a receiver located in (x_r, y_r, z_r) and a transmitter at (x_t, y_t, z_t) position, the range is given as:

$$r_{true} = c\, t_{propagation} = \sqrt{(x_t - x_r)^2 + (y_t - y_r)^2 + (z_t - z_r)^2} \qquad (6.4)$$

To solve Eq. 6.4 for (x_r, y_r, z_r), one requires ephemerides for three satellites. However, since the clock on the GPS transmitter (t_{GPS}) and the clock on the receiver (t_{local}) are not perfectly synchronized, there exists an offset t_r between these two time-scales. Therefore, the satellite-to-receiver distance that a receiver perceives is a pseudo-range (r_{pseudo}), where:

$$r_{pseudo} = r_{true} - c t_r \qquad (6.5)$$

Therefore, a GPS receiver needs to solve for four unknowns (x_r, y_r, z_r, t_r) for precise location and perfect synchronization. At least four satellite data is required for solving this system of equations. Using this solution, a GPS receiver updates its position, and synchronizes its local clock frequently to t_{sync} for keeping perfect synchronization with the universal coordinated time (UTC) where:

$$t_{sync} = t_{local} + t_r \qquad (6.6)$$

In practice, this method provides accuracy in the order of 10 meters in position and nearly 0.1 μs in time [32]. As a result, GPS signals can be used not only for positioning of the receivers but also for precise clock synchronization of receivers across the globe.

6.6.2   Hardware-Based Signal Authentication and Spoofing Detection

Here, we will demonstrate how to authenticate GPS time signals using intrinsic properties of a hardware oscillator. This would essentially construct a spoofing detector for the signal. This spoofing detector will calculate frequency states of this hardware clock using the received GPS signal as a reference. Any attacks on the received GPS data will create anomalies in the internal frequency states of this clock. Once an attack is detected, the design will attempt to generate an approximated version of the correct GPS time t_{GPS} to holdover the timing system during an attack. This design would require two additional resources in addition to the GPS receiver:

1. a single (or multiple) free-running oscillator(s),
2. additional data processing capabilities.

The architecture of the authentication system is shown in Fig. 6.7.

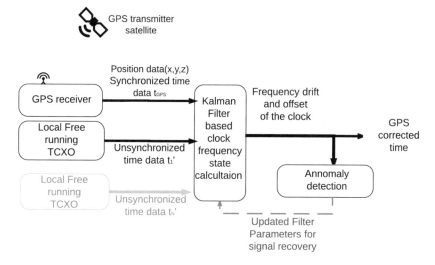

Fig. 6.7 Design of the secure GPS receiver with onboard authentication and spoofing detection mechanism. The receiver is equipped with a single (or multiple) free-running temperature controlled oscillator(s). Kalman filter-based state estimation is used for calculating frequency states of this clock. In the case of multiple clocks, time from each free-running oscillator can be used for generating a low noise stable virtual clock. Anomalies in the frequency drift and offset of this single clock (or the low noise virtual clock) calculated with respect to the GPS signal can reveal spoofing attacks on GPS signal. Updated filter parameters can help to reconstruct the approximated *true* time offset during a spoofing attack

Our approach depends on the internal frequency states (*i.e.*, frequency drifts and skew) of a hardware oscillator for spoofing detection. Hence, we provide details on hardware clock models in the next section to elucidate this approach.

6.6.3 Hardware Clocks

Clocks and oscillators in embedded systems are primarily used for time-keeping and synchronization purposes. In the majority of the embedded systems, crystal-based real-time clocks (RTCs) are used for precise time-keeping. These RTCs are far from perfect, and they deviate from ideal time due to both systematic and random variations. These systematic variations arise from the imperfections in the physical realization of the clock, and they are observed as time and frequency offsets and frequency drift.

At a given time t, the deviation of a clock from ideal time can be expressed as [33]:

$$x(t) = x_0 + y_0 t + \frac{1}{2} D t^2 + \epsilon(t) \tag{6.7}$$

where x_0, y_0, and D represents the time offset, frequency offset (also known as skew), and frequency drift. $\epsilon(t)$ represents non-deterministic random deviations. The frequency offsets and drifts of an RTC arise from the microscopic variations in the crystal used in these oscillators. The frequency offsets and drifts also vary with the dissimilarity in design, power supply, and environmental factors. These properties have been found to be different for similar oscillators working in a same operating condition. Therefore, we have the following assumptions regarding the frequency states of hardware clocks:

A1. Frequency drifts and skew of a clock with respect to a more precise reference are nearly constant and unique for a clock reference pair for a given duration.

A2. The states of a known free-running local oscillator are predictable for a given reference, and one can detect unusual activity in the reference by looking at the properties of the local oscillator.

A3. These properties vary uniquely for different clock pairs due to the random variations in their fabrication and are impossible to recreate without tampering the hardware of both the clocks.

Based on our assumption, the GPS induced internal states of a given free-running oscillator are relatively constant, and therefore, can be used to detect spoofing attacks. This is the key concept for our approach. It should be noted that Khono et al. [34] first proposed the idea of remote device fingerprinting using the uniqueness of frequency offset of hardware clocks. Since the publication of Khono's work [34], there has been a significant development in this field of remote device fingerprinting using hardware oscillators. Our assumptions can be validated by Khono's work, the subsequent works in the literature, and our experimental results and analysis presented in this work.

6.6.4 State Space Model of Hardware Clocks

For precisely calculating hardware clock states, we use a stochastic model of the clocks where a clock state is characterized by a column vector $x(t) = [x_1(t)\ y_1(t)\ D_1(t)]^T$. Here, $x_1(t)$, $y_1(t)$, and $D_1(t)$ represents the time offset state, frequency offset state, and frequency drift state, respectively. The clock state follows the stochastic difference equations as given in [35]:

$$\frac{dx_1}{dt} = y_1 + w_1;\ \frac{dy_1}{dt} = D_1 + w_2;\ \frac{dD_1}{dt} = w_3 \tag{6.8}$$

where, $w_i(t)$ represents the associated zero mean white noise with spectral densities q_i. For an ensemble of q clocks the state vector can be written as $[x_1, y_1, D_1 \ldots x_q, y_q, D_q]^T$. Discrete-time equations for a system described by 6.8 can be written as [35]:

$$\mathbf{X_n} = \mathbf{F_n X_{n-1}} + \mathbf{W_n} \tag{6.9}$$

$$\mathbf{\Xi_n} = \mathbf{H_n X_n} + \mathbf{V_n} \tag{6.10}$$

where, $n = 0, 1, 2, \ldots$ corresponds to discrete time t_n and measuring time interval $\Delta = t_n - t_{n-1}$.

For a single clock measurement, the $\mathbf{X_n} = [x_1, y_1, D_1]^T$ represents the state vector, and $\mathbf{\Xi_n}$ denotes the observation vector. $\mathbf{F_n}$ is the state transition matrix which is calculated as [35]:

$$\mathbf{F_n} = \begin{bmatrix} 1 & \Delta & \Delta^2/2 \\ 0 & 1 & \Delta \\ 0 & 0 & 1 \end{bmatrix} \tag{6.11}$$

The process noise $\mathbf{W_n}$ is considered to be zero mean additive white noise with covariance matrix $\mathbf{Q}$, where

$$\mathbf{Q} = \begin{bmatrix} q_1\Delta + q_2\frac{\Delta^3}{3} + q_3\frac{\Delta^5}{20} & q_2\frac{\Delta^2}{2} + q_3\frac{\Delta^4}{8} & q_3\frac{\Delta^3}{6} \\ q_2\frac{\Delta^2}{2} + q_3\frac{\Delta^4}{8} & q_2\Delta + q_3\frac{\Delta^3}{3} & q_3\frac{\Delta^2}{2} \\ q_3\frac{\Delta^3}{6} & q_3\frac{\Delta^2}{2} & q_3\Delta \end{bmatrix} \tag{6.12}$$

This state model for a clock ensemble is amenable to the design of optimal stochastic filters, which are broadly used for minimizing variance within a clock ensemble. In this work, we use this state space model for hardware oscillators and use an optimal filter (Kalman Formulation) to estimate these states for a single oscillator [35]. It should be noted that if there are more than one onboard free-running oscillators available, one could create a virtual time reference using an ensemble of clocks offering improved detection thresholds for spoofing attacks.

6.6.5 Kalman Filter Design for Authentication and Spoofing Detection

We use discrete-time state model for developing a Kalman filter-based spoofing detector. For our single clock experiment, we have the measurement matrix $\mathbf{H_n} = [1, 0, 0]$. $\mathbf{V_n}$ represents the zero mean measurement noise with covariance $\mathbf{R}$. For local measurements, we set $\mathbf{R} = \mathbf{0}$ to neglect the noise term. The algorithm for this linear Kalman filter [36] is given by the following equations:

Prediction Step:

$$\mathbf{m_{n|n-1}} = \mathbf{F_n}\mathbf{m_{n-1|n-1}} \tag{6.13}$$

$$\mathbf{P_{n|n-1}} = \mathbf{F_n}\mathbf{P_{n-1|n-1}}\mathbf{F_n^T} + \mathbf{Q} \tag{6.14}$$

Update Step:

$$\mathbf{K_n} = \mathbf{P_{n|n-1}}\mathbf{H_n^T}(\mathbf{H_n}\mathbf{P_{n|n-1}}\mathbf{H_n^T} + \mathbf{R})^{-1} \tag{6.15}$$

$$\mathbf{m_{n|n}} = \mathbf{m_{n|n-1}} + \mathbf{K_n}(\mathbf{\Xi_n} - \mathbf{H_n}\mathbf{m_{n|n-1}}) \tag{6.16}$$

$$\mathbf{P_{n|n}} = \mathbf{P_{n|n-1}} - \mathbf{K_n}\mathbf{H_n}\mathbf{P_{n|n-1}} \tag{6.17}$$

Here $\mathbf{m_{n|n}}$, $\mathbf{P_{n|n}}$ are the Gaussian posterior mean and covariance at nth time-step, and $\mathbf{K_n}$ is the Kalman gain at that step. The clock states at nth time-step is given by the components of $\mathbf{m_{n|n}}$, since $\mathbf{m_{n|n}}$ is the learned estimate of time offset, frequency offset, and drift at that step. For our simulations, we assume the Gaussian posterior mean at the beginning is zero and the initial posterior covariance is $\mathbb{I}^{3\times3}$. An example of calculating the time offset, frequency offset, and frequency drift for a crystal oscillator is shown in Fig. 6.8.

6.6.6 Signal Authentication and Anomaly Detection

Spoofing attacks induce variations in the GPS reference signal ranging from discrete step changes to slowly evolving changes in the demodulated GPS data. Our

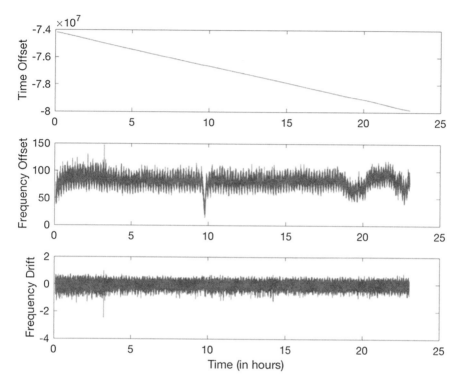

Fig. 6.8 Example time offset, frequency offset, and frequency drift for a crystal oscillator with respect to a stable GPS clock

anomaly detection strategy classifies changes in the time offset, frequency drift, and frequency offset measurements in the received GPS data in relation to the hardware oscillator as anomalous when $\mathbf{X_n}$ lies outside the confidence interval of its predicted value $\mathbf{m_{n|n-1}} \pm \mathbf{S_{n-1}}$, where,

$$\mathbf{S_{n-1}} = \mathbf{H_n P_{n|n-1} H_n^T} + \mathbf{R} \tag{6.18}$$

is the predicted variance of the offset. This approach can be used for detecting simpler attacks inducing a step change in the time offset. This technique depends only on a single data point and an estimate, and therefore, leads to a large number of false positives. Moreover, an *advanced* attacker will self-consistently change the offset to avoid such detection.

A better approach is to use a windowed strategy that takes account of a number of recent measurements and find out the likelihood of a new measurement and estimate. For this approach, we calculate

$$p(m_{i,n}|m_{i,n-1}, \ldots m_{i,n-k}) = \frac{1}{\sqrt{2\pi\sigma_{i,n-1}^2}} e^{\left(-\frac{(m_{i,n}-\bar{m}_{i,n-1})^2}{\sigma_{i,n-1}^2}\right)} \qquad (6.19)$$

where, $i \in \mathbb{Z}^+$ for a 3×1 Gaussian posterior mean, k is the window size, $\sigma_{i,n-1}^2$ is the variance, and $\bar{m}_{i,n-1}$ is the mean of the predicted values inside the window $(n-1)$ to $(n-k)$. By calculating a moving average of the log-likelihood z_n, we can detect an anomalous event when z_n crosses a predefined threshold. Here,

$$z_{i,n} = \alpha z_{i,n-1} + (1-\alpha)\ln(p(m_{i,n})) \qquad (6.20)$$

with α as the smoothing factor.

6.6.7   Spoofing Detection and Results

Adversary Model The major goal of the adversary is to produce erroneous time or position measurements in the GPS receiver. We assume that the adversary has complete access to the RF channel during the attack, *i.e.,* he can replay, alter, and/or replicate the RF carrier, spreading code and data bits of any or all of the visible satellites. We can divide the attacks in two categories:

1. Temporal shift injection
2. Meaconing and replay attack

A temporal shift injection attack changes the time and/or the position bits in the GPS signal, which is reflected as a sudden jump in the perceived time/location of the victim. A meaconing and replay attack first induces the receiver to lock onto its spoofed signal by transmitting code, phase, and Doppler-matched signals with gradually increasing power, and then drags off the code and phase carrier in such a way that he avoids a discontinuous step change in time or the location estimates of the victim receiver. We assume that the adversary is time-bound, *i.e.,* he has a limited time to spoof the receiver.

To demonstrate the proposed spoofing detection approach, first we consider an attacker performs a temporal shift injection attack on a GPS system. A temporal shift injection is initiated at 5000 seconds represented by a sudden jump in the offset as shown in Fig. 6.9. During the attack, the attacker maintains the temporal shift. When the attack ends, there is a sudden jump in the time offset which represents the recovery of the authentic signal by the receiver.

Detecting temporal shift injection in clock frequency domain is straightforward as one can notice the sudden overshoot in the estimated frequency offset and frequency drift. Therefore, simple jump detection techniques can be employed for discovering such attacks.

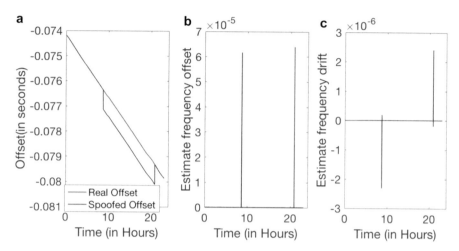

Fig. 6.9 Example of a temporal shift injection attack. (**a**) Clock offset of a free-running TCXO with respect to a GPS reference. A temporal shift injection attack is initiated at 5000 s. There is a sudden jump in the time offset which is corrected at the end of the attack; (**b**) Estimation of the frequency offset of the local clock with respect to the spoofed GPS signal during before and after the attack. (**c**) Similar estimation of frequency drift

For the meaconing and replay attack we assume that an attack scenario as described in [37]. The attack involves the deployment of a simulated gradually increasing delay on GPS signals, which results in an anomalous exaggeration of signal transmission time, and in turn, induces an offset error in the GPS receiver. This attack described by the authors has been used by other experimental evaluations of fault detection algorithms and provides a well-documented baseline to study the effectiveness of our proposed approach. Figure 6.10a illustrates the evolution of the attack starting at 5130 seconds causing a gradual deviation of the GPS trained clock (solid line) against the true reference (dashed line).

To detect the attack, we modeled the TCXOs using the stochastic model presented in Sect. 6.6.2. Existing time offset based data-level detection techniques can only detect spoofing if there is a discontinuous change in time, caused by a step change in the GPS reference. However, in this attack, the time is delayed slowly making the attack mostly undetectable. Since the process noise measurements of our TCXO were unknown, we used empirical values based on prior literature on clock jitter $q_1 = 10^{-3}, q_2 = 10^{-6}, q_3 = 10^{-9}$. We then design a state space model and use the Kalman filter formulation to detect anomalies in the received GPS signal.

From Fig. 6.10, we can see that if we use the simultaneous negative values of averaged log-likelihood of frequency drift and offset as an indication of spoofing attack, the detector can detect the first anomaly at 5752s, (about 10 minutes after the start of the attack). Note that in this particular experiment the spoofing attack was discovered when the cumulative error on the local clock was less than 4 µs. This is a relatively small error for some GPS-dependent systems.

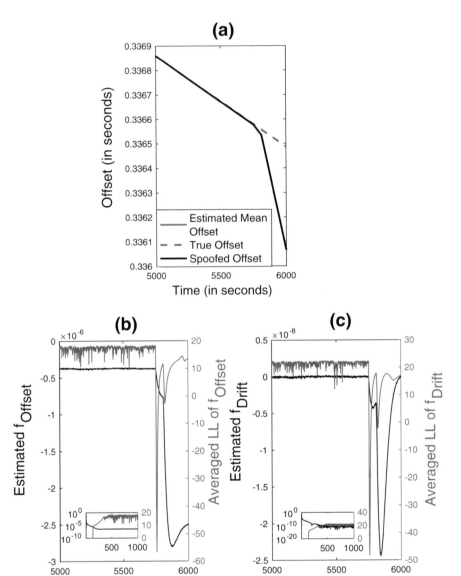

Fig. 6.10 Example of a replay and meaconing attack. (**a**) Clock offset of a free-running TCXO with respect to a GPS reference. A spoofing attack is initiated at 5130 seconds. The estimated time offset faithfully follows the spoofed signal as there is no sudden jump in the time offset; (**b**) Estimation of the frequency offset (black curve) and the averaged log-likelihood of the frequency offset (red curve) of the local clock with respect to the spoofed GPS signal during before and after the attack. The initial start-up and transition period for f_{offset} and the log-likelihood is shown in the inset. (**c**) Similar estimation of frequency drift (black curve) and the averaged log-likelihood of the frequency drift (red curve) of the local oscillator. The window size (k) for this computation is 128 data points

6.6.8 Analysis of Hardware Dependent Signal Authentication

Accuracy The accuracy of the proposed detection technique depends on the noise margin and stability of the hardware clocks. For our experiments, we have used inexpensive temperature controlled crystal oscillators (TCXOs), which provides a degree of immunity to temperature variations. One can use an ensemble of hardware oscillators to create a virtual clock using the system of equations as discussed in Sect. 6.6.3 to reduce this noise. It should be noted that the accuracy and robustness of our detection mechanism require a prior estimate of the measurement noise **Q**.

Computation Cost The cost of matrix-vector computations for a Kalman filter in the prediction and update step contains computation in the order of $O(\mathbf{D}^2)$, $O(\mathbf{MD})$, and $O(\mathbf{M}^3)$ complexity. The covariance matrices are symmetric, and therefore, Cholesky factorization can be used for maintaining **P** in a square-root form. Since the prediction and update step requires the knowledge of only current and previous steps, this construction has a very low memory complexity. The anomaly detection step has logarithmic complexity which can be simplified by approximating $p(m_{i,n}) \approx e^{\left(-\frac{(m_{i,n}-\bar{m}_{i,n-1})^2}{\sigma_{i,n-1}^2}\right)}$. The averaging window has a fixed memory requirement which can be lowered by reducing the number of historical data points.

Hardware Overhead GPS receivers already contain a hardware oscillator which is synchronized using the GPS signals. By turning off the synchronization, it may be possible to convert this clock to a free-running oscillator. The synchronization based timing corrections can be performed in software. Another approach is to add a hardware component with embedded free-running oscillators to employ the proposed method without altering the GPS receiver design. The computation can be performed using onboard processors in IoT devices or by adding a low power microcontroller that takes GPS-derived time as an input and provides the corrected time and spoofing detection capability to the system.

In summary, we present a design for integrating data-level spoofing detection with an existing GPS-based timing system in this section. The design uses single (or multiple) free-running oscillators to detect anomalies in the GPS-derived frequency drift and the offset. We demonstrate that this approach can provide fast and accurate detection of GPS spoofing attacks published in the literature. Since GPS spoofing attacks pose a significant threat to IoT systems, including spoofing detection methods such as the method presented in this chapter, in future GPS receiver designs will secure future navigation hardware for IoT application.

6.7 Conclusions

In this chapter, we have demonstrated several examples of lightweight authentication using hardware dependent techniques. As the IoT space becomes larger, new and efficient security protocols will be required to support a wide, distributed low power networks. Novel methods ensuring security and privacy will be necessary, as well as existing cryptographic techniques need to be revisited for this purpose. From a hardware engineering point of view—when power and area budget become crucial, techniques similar to the ones discussed in this work will be cost effective and energy efficient.

References

1. G. Qu, L. Yuan, Design things for the internet of things—an EDA perspective, in *2014 IEEE/ACM International Conference on Computer-Aided Design (ICCAD)* (IEEE, 2014), pp. 411–416
2. A.J. Menezes, P.C. Van Oorschot, S.A. Vanstone, *Handbook of Applied Cryptography* (CRC press, 1996)
3. M.T. Arafin, G. Qu, RRAM based lightweight user authentication, in *Proceedings of the IEEE/ACM International Conference on Computer-Aided Design, ser. ICCAD '15* (IEEE Press, Piscataway, NJ, USA, 2015), pp. 139–145. [Online]. Available: http://dl.acm.org/citation.cfm?id=2840819.2840839
4. A. Shamir, How to share a secret. Commun. ACM **22**(11), 612–613 (1979)
5. G.R. Blakley, Safeguarding cryptographic keys, in *Proc. of the National Computer Conference 1979*, vol. 48, pp. 313–317, 1979
6. D.R. Stinson, An explication of secret sharing schemes. Des. Codes Crypt. **2**(4), 357–390 (1992)
7. M. Naor, A. Shamir, Visual cryptography, in *Workshop on the Theory and Application of Cryptographic Techniques* (Springer, 1994), pp. 1–12
8. N. Beckmann, M. Potkonjak, Hardware-based public-key cryptography with public physically unclonable functions, in *International Workshop on Information Hiding* (Springer, 2009), pp. 206–220
9. R. Maes, *Physically Unclonable Functions* (Springer, 2013)
10. R. Maes, I. Verbauwhede, Physically unclonable functions: A study on the state of the art and future research directions, in *Towards Hardware-Intrinsic Security* (Springer, 2010), pp. 3–37
11. G.E. Suh, S. Devadas, Physical unclonable functions for device authentication and secret key generation, in *Proceedings of the 44th annual Design Automation Conference* (ACM, 2007), pp. 9–14
12. M. Naor, B. Pinkas, Visual authentication and identification, in *Annual International Cryptology Conference* (Springer, 1997), pp. 322–336
13. M. Elgebaly, M. Sachdev, Variation-aware adaptive voltage scaling system. IEEE Trans. Very Large Scale Integr. VLSI Syst. **15**(5), 560–571 (2007)
14. N. Banerjee, G. Karakonstantis, K. Roy, Process variation tolerant low power DCT architecture, in *Proceedings of the Conference on Design, Automation and Test in Europe* (EDA Consortium, 2007), pp. 630–635
15. J. Han, M. Orshansky, Approximate computing: An emerging paradigm for energy-efficient design, in *2013 18th IEEE European Test Symposium (ETS)* (IEEE, 2013), pp. 1–6
16. R. Venkatesan, A. Agarwal, K. Roy, A. Raghunathan, Macaco: Modeling and analysis of circuits for approximate computing, in *Proceedings of the International Conference on Computer-Aided Design* (IEEE Press, 2011), pp. 667–673
17. F. Ganji, S. Tajik, J.-P. Seifert, Why attackers win: on the learnability of XOR arbiter PUFS, in *International Conference on Trust and Trustworthy Computing* (Springer, 2015), pp. 22–39

18. J.E. Stine, I. Castellanos, M. Wood, J. Henson, F. Love, W.R. Davis, P.D. Franzon, M. Bucher, S. Basavarajaiah, J. Oh et al., FreePDK: An open-source variation-aware design kit, in *2007 IEEE International Conference on Microelectronic Systems Education (MSE'07)* (IEEE, 2007), pp. 173–174

19. M.T. Arafin, M. Gao, G. Qu, Volta: Voltage over-scaling based lightweight authentication for IoT applications, in *2017 22nd Asia and South Pacific Design Automation Conference (ASP-DAC)* (IEEE, 2017), pp. 336–341

20. J. Rajendran, G.S. Rose, R. Karri, M. Potkonjak, Nano-PPUF: A memristor-based security primitive, in *Computer Society Annual Symposium on VLSI (ISVLSI)* (IEEE, 2012), pp. 84–87

21. J.B. Wendt, M. Potkonjak, The bidirectional polyomino partitioned PPUF as a hardware security primitive, in *Global Conference on Signal and Information Processing (GlobalSIP)* (IEEE, 2013), pp. 257–260

22. J. Rajendran, R. Karri, J.B. Wendt, M. Potkonjak, N.R. McDonald, G.S. Rose, B.T. Wysocki, Nanoelectronic solutions for hardware security, in *IACR Cryptology ePrint Archive*, vol. 2012, p. 575, 2012

23. H. Kim, M.P. Sah, C. Yang, L.O. Chua, Memristor-based multilevel memory, in *2010 12th International Workshop on Cellular Nanoscale Networks and Their Applications (CNNA)* (IEEE, 2010), pp. 1–6

24. Y. Dodis, L. Reyzin, A. Smith, Fuzzy extractors: How to generate strong keys from biometrics and other noisy data, in *International Conference on the Theory and Applications of Cryptographic Techniques* (Springer, 2004), pp. 523–540

25. G. Horng, T. Chen, D.-S. Tsai, Cheating in visual cryptography, *Designs, Codes and Cryptography*, vol. 38(2), pp. 219–236, 2006

26. C. Bonebrake, L.R. O'Neil, Attacks on GPS time reliability. IEEE Secur. Priv. **12**(3), 82–84 (2014)

27. T.E. Humphreys, B.M. Ledvina, M.L. Psiaki, B.W. OâHanlon, P.M. Kintner Jr, Assessing the spoofing threat: development of a portable GPS civilian spoofer, in *Proceedings of the ION GNSS International Technical Meeting of the Satellite Division*, vol. 55, p. 56, 2008

28. A. Jafarnia-Jahromi, A. Broumandan, J. Nielsen, G. Lachapelle, GPS vulnerability to spoofing threats and a review of antispoofing techniques. Int. J. Navig. Obs. **2012**,* (2012). https://doi.org/10.1155/2012/127072

29. M.G. Kuhn, Signal authentication in trusted satellite navigation receivers, in *Towards Hardware-Intrinsic Security* (Springer, 2010), pp. 331–348

30. N.O. Tippenhauer, C. Pöpper, K.B. Rasmussen, S. Capkun, On the requirements for successful GPS spoofing attacks, in *Proceedings of the 18th ACM Conference on Computer and Communications Security* (ACM, 2011), pp. 75–86

31. P. Misra, P. Enge, *Global Positioning System: Signals, Measurements and Performance Second Edition* (Ganga-Jamuna Press, Lincoln, MA, 2006)

32. P.H. Dana, B.M. Penrod, The role of GPS in precise time and frequency dissemination, in *GPS World*, pp. 38–43, 1990

33. D.W. Allan, Time and frequency(time-domain) characterization, estimation, and prediction of precision clocks and oscillators. IEEE Trans. Ultras. Ferroelectr. Freq. Control **34**(6), 647–654 (1987)

34. T. Kohno, A. Broido, K.C. Claffy, Remote physical device fingerprinting. IEEE Trans. Dependable Secure Comput. **2**(2), 93–108 (2005)

35. C.A. Greenhall, A review of reduced Kalman filters for clock ensembles. IEEE Trans. Ultras. Ferroelectr. Freq. Control **59**(3), 491–496 (2012)

36. S.S. Haykin, *Adaptive Filter Theory* (Pearson Education India, 2008)

37. D.P. Shepard, T.E. Humphreys, A.A. Fansler, Evaluation of the vulnerability of phasor measurement units to GPS spoofing attacks. Int. J. Crit. Infrastruct. Prot. **5**(3), 146–153 (2012)

Index

© Springer Nature Switzerland AG 2021
B. Halak (ed.), *Authentication of Embedded Devices*,
https://doi.org/10.1007/978-3-030-60769-2